Statistical Distributions

Statistical Distributions

Fourth Edition

Catherine Forbes
Monash University, Victoria, Australia

Merran Evans
Monash University, Victoria, Australia

Nicholas Hastings
Albany Interactive, Victoria, Australia

Brian Peacock
Brian Peacock Ergonomics, SIM University, Singapore

WILEY
A JOHN WILEY & SONS, INC., PUBLICATION

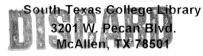

Published by John Wiley & Sons, Inc., Hoboken, New Jersey.
Published simultaneously in Canada.

Library of Congress Cataloging-in-Publication Data:

Statistical distributions. – 4th ed. / Catherine Forbes ... [et al.].
p. cm. – (Wiley series in probability and statistics)
Includes bibliographical references and index.
ISBN 978-0-470-39063-4 (pbk.)
1. Distribution (Probability theory) I. Forbes, Catherine.
QA273.6.E92 2010
519.2'4–dc22

2009052131

Printed in the United States of America.

10 9 8 7 6 5 4 3 2 1

TO

Jeremy and Elana Forbes

Caitlin and Eamon Evans

Tina Hastings

Eileen Peacock

Contents

Preface

This revised handbook provides a concise summary of the salient facts and formulas relating to 40 major probability distributions, together with associated diagrams that allow the shape and other general properties of each distribution to be readily appreciated.

In the introductory chapters the fundamental concepts of the subject are covered with clarity, and the rules governing the relationships between variates are described. Extensive use is made of the inverse distribution function and a definition establishes a variate as a generalized form of a random variable. A consistent and unambiguous system of nomenclature can thus be developed, with chapter summaries relating to individual distributions.

Students, teachers, and practitioners for whom statistics is either a primary or secondary discipline will find this book of great value, both for factual references and as a guide to the basic principles of the subject. It fulfills the need for rapid access to information that must otherwise be gleaned from many scattered sources.

The first version of this book, written by N. A. J. Hastings and J. B. Peacock, was published by Butterworths, London, 1975. The second edition, with a new author, M. A. Evans, was published by John Wiley & Sons in 1993, with a third edition by the same authors published by John Wiley & Sons in 2000. This fourth edition sees the addition of a new author, C. S. Forbes. Catherine Forbes holds a Ph.D. in Mathematical Statistics from The Ohio State University, USA, and is currently Senior Lecturer at Monash University, Victoria, Australia. Professor Merran Evans is currently Pro Vice-Chancellor, Planning and Quality at Monash University and obtained her Ph.D. in Econometrics from Monash University. Dr. Nicholas Hastings holds a Ph.D. in Operations Research from the University of Birmingham. Formerly Mount Isa Mines Professor of Maintenance Engineering at Queensland University of Technology, Brisbane, Australia, he is currently Director and Consultant in physical asset management, Albany Interactive Pty Ltd. Dr. Brian Peacock has a background in ergonomics and industrial engineering which have provided a foundation for a long career in industry and academia, including 18 years in academia, 15 years with General Motors' vehicle design and manufacturing organizations, and 4 years as discipline coordinating scientist for the National Space Biomedical Institute/NASA. He is a licensed professional engineer, a licensed private pilot, a certified professional ergonomist, and a fellow of both the Ergonomics and Human Factors Society (UK) and the Human Factors and Ergonomics Society (USA). He recently retired as a professor in the Department of Safety Science at Embry Riddle Aeronautical University, where he taught classes in system safety and applied ergonomics.

The authors gratefully acknowledge the helpful suggestions and comments made by Harry Bartlett, Jim Conlan, Benoit Dulong, Alan Farley, Robert Kushler, Jerry W. Lewis, Allan T. Mense, Grant Reinman, and Dimitris Ververidis.

CATHERINE FORBES
MERRAN EVANS
NICHOLAS HASTINGS
BRIAN PEACOCK

Chapter 1

Introduction

The number of puppies in a litter, the life of a light bulb, and the time to arrival of the next bus at a stop are all examples of random variables encountered in everyday life. Random variables have come to play an important role in nearly every field of study: in physics, chemistry, and engineering, and especially in the biological, social, and management sciences. Random variables are measured and analyzed in terms of their statistical and probabilistic properties, an underlying feature of which is the distribution function. Although the number of potential distribution models is very large, in practice a relatively small number have come to prominence, either because they have desirable mathematical characteristics or because they relate particularly well to some slice of reality or both.

This book gives a concise statement of leading facts relating to 40 distributions and includes diagrams so that shapes and other general properties may readily be appreciated. A consistent system of nomenclature is used throughout. We have found ourselves in need of just such a summary on frequent occasions—as students, as teachers, and as practitioners. This book has been prepared and revised in an attempt to fill the need for rapid access to information that must otherwise be gleaned from scattered and individually costly sources.

In choosing the material, we have been guided by a utilitarian outlook. For example, some distributions that are special cases of more general families are given extended treatment where this is felt to be justified by applications. A general discussion of families or systems of distributions was considered beyond the scope of this book. In choosing the appropriate symbols and parameters for the description of each distribution, and especially where different but interrelated sets of symbols are in use in different fields, we have tried to strike a balance between the various usages, the need for a consistent system of nomenclature within the book, and typographic simplicity. We have given some methods of parameter estimation where we felt it was appropriate to do so. References listed in the Bibliography are not the primary sources but should be regarded as the first "port of call".

In addition to listing the properties of individual variates we have considered relationships between variates. This area is often obscure to the nonspecialist. We

Statistical Distributions, Fourth Edition, by Catherine Forbes, Merran Evans, Nicholas Hastings, and Brian Peacock

have also made use of the inverse distribution function, a function that is widely tabulated and used but rarely explicitly defined. We have particularly sought to avoid the confusion that can result from using a single symbol to mean here a function, there a quantile, and elsewhere a variate.

Building on the three previous editions, this fourth edition documents recent extensions to many of these probability distributions, facilitating their use in more varied applications. Details regarding the connection between joint, marginal, and conditional probabilities have been included, as well as new chapters (Chapters 5 and 6) covering the concepts of statistical modeling and parameter inference. In addition, a new chapter (Chapter 38) detailing many of the existing standard queuing theory results is included. We hope the new material will encourage readers to explore new ways to work with statistical distributions.

Chapter 2

Terms and Symbols

2.1 PROBABILITY, RANDOM VARIABLE, VARIATE, AND NUMBER

Probabilistic Experiment

A probabilistic experiment is some occurrence such as the tossing of coins, rolling dice, or observation of rainfall on a particular day where a complex natural background leads to a chance outcome.

Sample Space

The set of possible outcomes of a probabilistic experiment is called the sample, event, or possibility space. For example, if two coins are tossed, the sample space is the set of possible results HH, HT, TH, and TT, where H indicates a head and T a tail.

Random Variable

A random variable is a function that maps events defined on a sample space into a set of values. Several different random variables may be defined in relation to a given experiment. Thus, in the case of tossing two coins the number of heads observed is one random variable, the number of tails is another, and the number of double heads is another. The random variable "number of heads" associates the number 0 with the event TT, the number 1 with the events TH and HT, and the number 2 with the event HH. Figure 2.1 illustrates this mapping.

Variate

In the discussion of statistical distributions it is convenient to work in terms of variates. A variate is a generalization of the idea of a random variable and has similar

Statistical Distributions, Fourth Edition, by Catherine Forbes, Merran Evans, Nicholas Hastings, and Brian Peacock

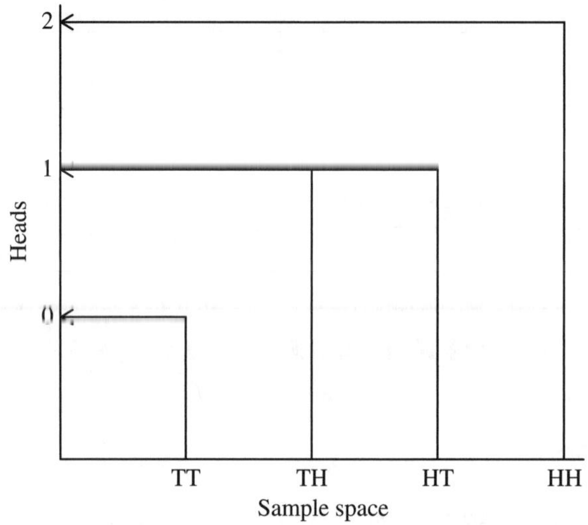

Figure 2.1. The random variable "number of heads".

probabilistic properties but is defined without reference to a particular type of probabilistic experiment. A *variate* is the set of all random variables that obey a given probabilistic law. The number of heads and the number of tails observed in independent coin tossing experiments are elements of the same variate since the probabilistic factors governing the numerical part of their outcome are identical.

A *multivariate* is a vector or a set of elements, each of which is a variate. A *matrix variate* is a matrix or two-dimensional array of elements, each of which is a variate. In general, dependencies may exist between these elements.

Random Number

A *random number* associated with a given variate is a number generated at a realization of any random variable that is an element of that variate.

2.2 RANGE, QUANTILE, PROBABILITY STATEMENT, AND DOMAIN

Range

Let X denote a variate and let $\Re_X$ be the set of all (real number) values that the variate can take. The set $\Re_X$ is the *range* of X. As an illustration (illustrations are in terms of random variables) consider the experiment of tossing two coins and noting the number of heads. The range of this random variable is the set $\{0, 1, 2\}$ heads, since the result may show zero, one, or two heads. (An alternative common usage of the term *range* refers to the largest minus the smallest of a set of variate values.)

Quantile

For a general variate X let x (a real number) denote a general element of the range $\Re_X$. We refer to x as the *quantile* of X. In the coin tossing experiment referred to previously, $x \in \{0, 1, 2\}$ heads; that is, x is a member of the set $\{0, 1, 2\}$ heads.

Probability Statement

Let $X = x$ mean "the value realized by the variate X is x." Let $\Pr[X \leq x]$ mean "the probability that the value realized by the variate X is less than or equal to x."

Probability Domain

Let α (a real number between 0 and 1) denote probability. Let $\Re_X^\alpha$ be the set of all values (of probability) that $\Pr[X \leq x]$ can take. For a continuous variate, $\Re_X^\alpha$ is the line segment [0, 1]; for a discrete variate it will be a subset of that segment. Thus $\Re_X^\alpha$ is the *probability domain* of the variate X.

In examples we shall use the symbol X to denote a random variable. Let X be the number of heads observed when two coins are tossed. We then have

$$\Pr[X \leq 0] = \frac{1}{4}$$

$$\Pr[X \leq 1] = \frac{3}{4}$$

$$\Pr[X \leq 2] = 1$$

and hence $\Re_X^\alpha = \{\frac{1}{4}, \frac{3}{4}, 1\}$.

2.3 DISTRIBUTION FUNCTION AND SURVIVAL FUNCTION

Distribution Function

The *distribution function* F (or more specifically F_X) associated with a variate X maps from the range $\Re_X$ into the probability domain $\Re_X^\alpha$ or [0, 1] and is such that

$$F(x) = \Pr[X \leq x] = \alpha \qquad x \in \Re_X, \alpha \in \Re_X^\alpha. \tag{2.1}$$

The function $F(x)$ is nondecreasing in x and attains the value unity at the maximum of x. Figure 2.2 illustrates the distribution function for the number of heads in the experiment of tossing two coins. Figure 2.3 illustrates a general continuous distribution function and Figure 2.4 a general discrete distribution function.

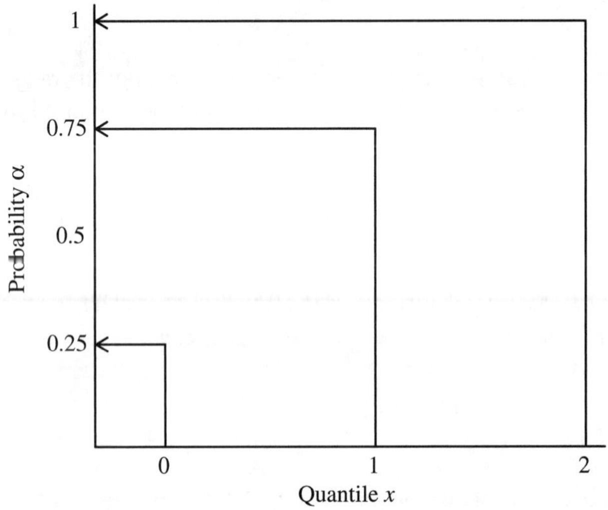

Figure 2.2. The distribution function $F: x \to \alpha$ or $\alpha = F(x)$ for the random variable, "number of heads".

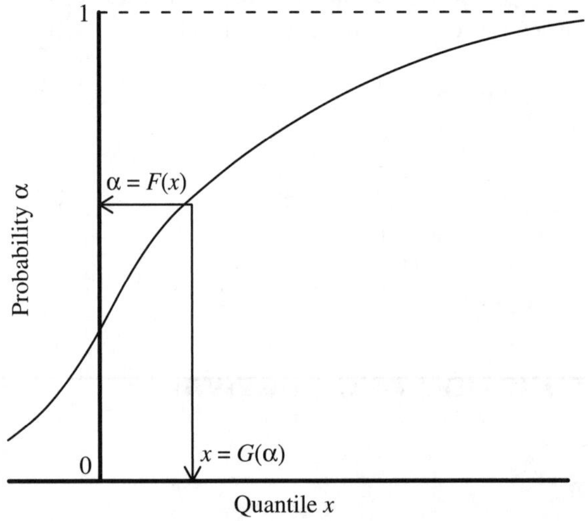

Figure 2.3. Distribution function and inverse distribution function for a continuous variate.

Survival Function

The *survival function* $S(x)$ is such that

$$S(x) = \Pr[X > x] = 1 - F(x).$$

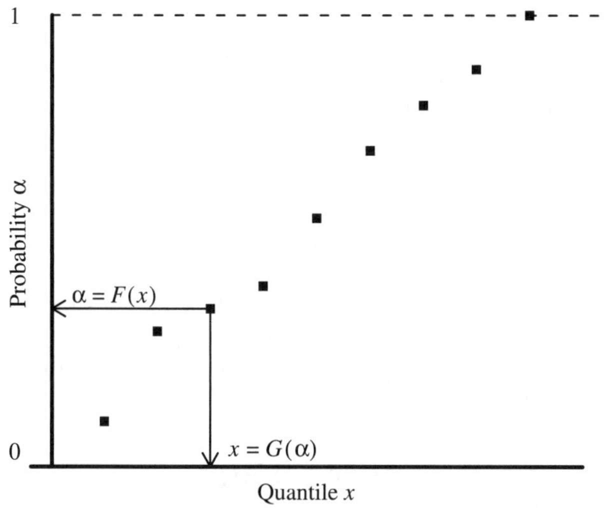

Figure 2.4. Distribution function and inverse distribution function for a discrete variate.

2.4 INVERSE DISTRIBUTION FUNCTION AND INVERSE SURVIVAL FUNCTION

For a distribution function F, mapping a quantile x into a probability α, the quantile function or inverse distribution function G performs the corresponding inverse mapping from α into x. Thus for $x \in \Re_X$, $\alpha \in \Re_X^\alpha$, the following statements hold:

$$\alpha = F(x) \tag{2.2}$$

$$x = G(\alpha) \tag{2.3}$$

$$x = G(F(x))$$

$$\alpha = F(G(\alpha))$$

$$\Pr[X \leq x] = F(x) = \alpha$$

$$\Pr[X \leq G(\alpha)] = F(x) = \alpha \tag{2.4}$$

where $G(\alpha)$ is the quantile such that the probability that the variate takes a value less than or equal to it is α; $G(\alpha)$ is the 100α percentile.

Figures 2.2, 2.3, and 2.4 illustrate both distribution functions and inverse distribution functions, the difference lying only in the choice of independent variable.

For the two-coin tossing experiment the distribution function F and inverse distribution function G of the number of heads are as follows:

$$F(0) = \tfrac{1}{4} \quad G\left(\tfrac{1}{4}\right) = 0$$

$$F(1) = \tfrac{3}{4} \quad G\left(\tfrac{3}{4}\right) = 1$$

$$F(2) = 1 \quad G(1) = 2$$

Inverse Survival Function

The inverse survival function Z is a function such that $Z(\alpha)$ is the quantile, which is exceeded with probability α. This definition leads to the following equations:

$$\Pr[X > Z(\alpha)] = \alpha$$

$$Z(\alpha) = G(1 - \alpha)$$

$$x = Z(\alpha) = Z(S(x))$$

Inverse survival functions are among the more widely tabulated functions in statistics. For example, the well-known chi-squared tables are tables of the quantile x as a function of the probability level α and a shape parameter, and hence are tables of the chi-squared inverse survival function.

2.5 PROBABILITY DENSITY FUNCTION AND PROBABILITY FUNCTION

A probability density function, $f(x)$, is the first derivative coefficient of a distribution function, $F(x)$, with respect to x (where this derivative exists).

$$f(x) = \frac{d(F(x))}{dx}$$

For a given continuous variate $\boldsymbol{X}$ the area under the probability density curve between two points x_L, x_U in the range of $\boldsymbol{X}$ is equal to the probability that an as-yet unrealized random number of $\boldsymbol{X}$ will lie between x_L and x_U. Figure 2.5 illustrates this. Figure 2.6

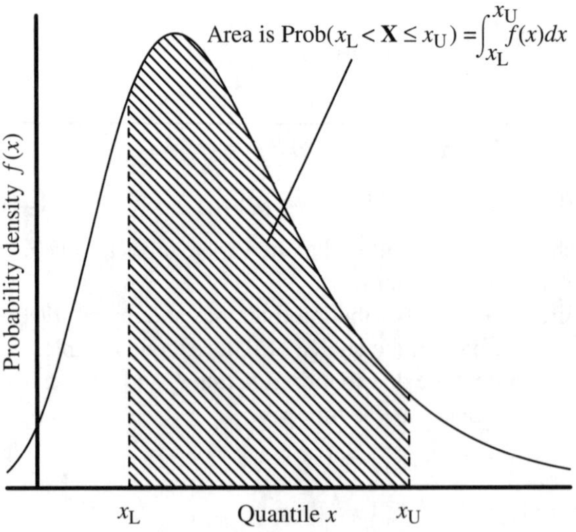

Figure 2.5. Probability density function.

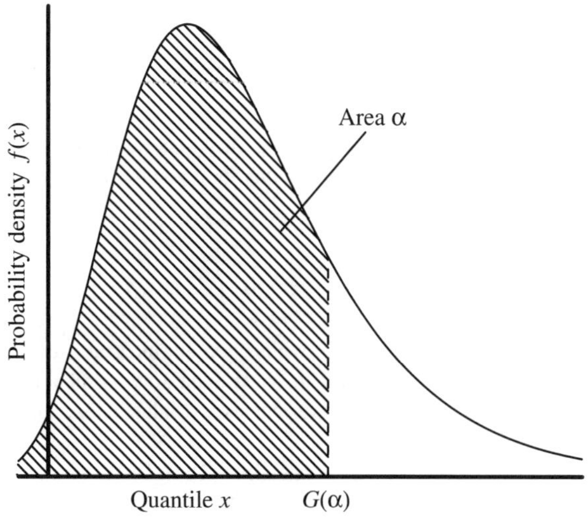

Figure 2.6. Probability density function illustrating the quantile corresponding to a given probability α; G is the inverse distribution function.

illustrates the relationship between the area under a probability density curve and the quantile mapped by the inverse distribution function at the corresponding probability value.

A discrete variate takes discrete values x with finite probabilities $f(x)$. In this case $f(x)$ is the probability function, also called the probability mass function.

2.6 OTHER ASSOCIATED FUNCTIONS AND QUANTITIES

In addition to the functions just described, there are many other functions and quantities that are associated with a given variate. A listing is given in Table 2.1 relating to a general variate that may be either continuous or discrete. The integrals in Table 2.1 are Stieltjes integrals, which for discrete variates become ordinary summations, so

$$\int_{x_L}^{x_U} \phi(x) f(x) dx, \text{ corresponds to } \sum_{x=x_L}^{x_U} \phi(x) f(x).$$

Table 2.2 gives some general relationships between moments, and Table 2.3 gives our notation for values, mean, and variance for samples.

Table 2.1: Functions and Related Quantities for a General Variate (**X** Denotes a Variate, x a Quantile, and α a Probability)

Term	Symbol	Description and notes
1. Distribution function (df) or cumulative distribution function (cdf)	$F(x)$	$F(x)$ is the probability that the variate takes a value less than or equal to x. $$F(x) = \Pr[X \le x] = \alpha$$ $$F(x) = \int_{-\infty}^{x} f(u)\,du$$
2. Probability density function (pdf) (continuous variates)	$f(x)$	A function whose general integral over the range x_L to x_U is equal to the probability that the variate takes a value in that range. $$\int_{x_L}^{x_U} f(x)\,dx = \Pr[x_L < X \le x_U]$$ $$f(x) = \frac{d(F(x))}{dx}$$
3. Probability function (pf) (discrete variates)	$f(x)$	$f(x)$ is the probability that the variate takes the value x. $$f(x) = \Pr[X = x]$$
4. Inverse distribution function or quantile function (of probability α)	$G(\alpha)$	$G(\alpha)$ is the quantile such that the probability that the variate takes a value less than or equal to it is α. $$x = G(\alpha) = G(F(x))$$ $$\Pr[X \le G(\alpha)] = \alpha$$ $G(\alpha)$ is the $100\,\alpha$ percentile. The relation to df and pdf is shown in Figures 2.3, 2.4, and 2.6.
5. Survival function	$S(x)$	$S(x)$ is the probability that the variate takes a value greater than x. $$S(x) = \Pr[X > x] = 1 - F(x)$$
6. Inverse survival function (of probability α)	$Z(\alpha)$	$Z(\alpha)$ is the quantile that is exceeded by the variate with probability α. $$\Pr[X > Z(\alpha)] = \alpha$$ $$x = Z(\alpha) = Z(S(x))$$ where S is the survival function $$Z(\alpha) = G(1 - \alpha)$$ and G is the inverse distribution function.

Table 2.1: (*Continued*)

Term	Symbol	Description and notes
7. Hazard function (or failure rate, hazard rate, or force of mortality)	$h(x)$	$h(x)$ is the ratio of the probability density function to the survival function at quantile x. $$h(x) = f(x)/S(x)$$ $$h(x) = f(x)/(1 - F(x))$$
8. Mills ratio	$m(x)$	$m(x)$ is the inverse of the hazard function. $$m(x) = (1 - F(x))/f(x) = 1/h(x)$$
9. Cumulative or integrated hazard function	$H(x)$	Integral of the hazard function. $$H(x) = \int_{-\infty}^{x} h(u)du$$ $$H(x) = -\log(1 - F(x))$$ $$S(x) = 1 - F(x) = \exp(-H(x))$$
10. Probability generating function (discrete nonnegative integer valued variates); also called the geometric or z transform	$P(t)$	A function of an auxiliary variable t (or alternatively z) such that the coefficient of $t^x = f(x)$. $$P(t) = \sum_{x=0}^{\infty} t^x f(x)$$ $$f(x) = \left(\frac{1}{x!}\right)\left(\frac{d^x P(t)}{dt^x}\right)_{t=0}$$
11. Moment generating function (mgf)	$M(t)$	A function of an auxiliary variable t whose general term is of the form $\mu'_r t^r/r!$. $$M(t) = \int_{-\infty}^{\infty} \exp(tx)f(x)dx$$ $$M(t) = 1 + \mu'_1 t + \mu'_2 t^2/2! + \cdots + \mu'_r t^r/r! + \cdots$$ For any independent variates $\boldsymbol{A}$ and $\boldsymbol{B}$, whose moment generating functions $M_{\boldsymbol{A}}(t)$ and $M_{\boldsymbol{B}}(t)$ respectively, exist, then the mgf of $\mathbf{A} + \mathbf{B}$ exists and satisfies $$M_{\boldsymbol{A}+\boldsymbol{B}}(t) = M_{\boldsymbol{A}}(t)M_{\boldsymbol{B}}(t).$$

(*continued*)

Table 2.1: (*Continued*)

Term	Symbol	Description and notes
12. Laplace transform of the pdf	$f^*(s)$	A function of the auxiliary variable s defined by $$f^*(s) = \int_0^{\infty} \exp(-sx) f(x)dx, \quad x \geq 0.$$
13. Characteristic function	$C(t)$	A function of the auxiliary variable t and the imaginary quantity i ($i^2 = -1$), which exists and is unique to a given pdf. $$C(t) = \int_{-\infty}^{+\infty} \exp(itx) f(x)dx$$ If $C(t)$ is expanded in powers of t and if μ'_r exists, then the general term is $\mu'_r(it)^r/r!$ For any independent variates $\boldsymbol{A}$ and $\boldsymbol{B}$ $$C_{\boldsymbol{A}+\boldsymbol{B}}(t) = C_{\boldsymbol{A}}(t)C_{\boldsymbol{B}}(t)$$
14. Cumulant generation function	$K(t)$	$K(t) = \log C(t)$ [sometimes defined $\log M(t)$]. $$K_{\boldsymbol{A}+\boldsymbol{B}}(t) = K_{\boldsymbol{A}}(t) + K_{\boldsymbol{B}}(t)$$
15. rth Cumulant	k_r	The coefficient of $(it)^r/r!$ in the expansion of $K(t)$.
16. rth Moment about the origin	μ'_r	$$\mu'_r = \int_{-\infty}^{\infty} x^r f(x)dx$$ $$\mu'_r = \left(\frac{d^r M(t)}{dt^r}\right)_{t=0}$$ $$\mu'_r = (-i)^r \left(\frac{d^r C(t)}{dt^r}\right)_{t=0}$$
17. Mean (first moment about the origin)	μ	$$\mu = \int_{-\infty}^{+\infty} x f(x)dx = \mu'_1$$
18. rth (Central) moment about the mean	μ_r	$$\mu_r = \int_{-\infty}^{+\infty} (x-\mu)^r f(x)dx$$
19. Variance (second moment about the mean, μ_2)	σ^2	$$\sigma^2 = \int_{-\infty}^{+\infty} (x-\mu)^2 f(x)dx$$ $$= \mu_2 = \mu'_2 - \mu^2$$

Table 2.1: (*Continued*)

Term	Symbol	Description and notes
20. Standard deviation	σ	The positive square root of the variance.
21. Mean deviation		$\int_{-\infty}^{+\infty} \lvert x-\mu \rvert f(x)dx$. The mean absolute value of the deviation from the mean.
22. Mode		A quantile for which the pdf or pf is a local maximum.
23. Median	m	The quantile that is exceeded with probability $\frac{1}{2}$, $m = G(\frac{1}{2})$.
24. Quartiles		The upper and lower quartiles are exceeded with probabilities $\frac{1}{4}$ and $\frac{3}{4}$, corresponding to $G(\frac{1}{4})$ and $G(\frac{3}{4})$, respectively.
25. Percentiles		$G(\alpha)$ is the $100\,\alpha$ percentile.
26. Standardized rth moment about the mean	η_r	The rth moment about the mean scaled so that the standard deviation is unity. $\eta_r = \int_{-\infty}^{+\infty} \left(\frac{x-\mu}{\sigma}\right)^r f(x)dx = \frac{\mu_r}{\sigma_r}$
27. Coefficient of skewness	η_3	$\sqrt{\beta_1} = \eta_3 = \mu_3/\sigma^3 = \mu_3/\mu_2^{3/2}$
28. Coefficient of kurtosis	η_4	$\beta_2 = \eta_4 = \mu_4/\sigma^4 = \mu_4/\mu_2^2$ Coefficient of excess or excess kurtosis is $\beta_2 - 3$; $\beta_2<3$ is platykurtosis; $\beta_2 > 3$ is leptokurtosis.
29. Coefficient of variation		Standard deviation/mean $= \sigma/\mu$.
30. Information content (or entropy)	I	$I = -\int_{-\infty}^{+\infty} f(x)\log_2(f(x))dx$
31. rth Factorial moment about the origin (discrete nonnegative variates)	$\mu'_{(r)}$	$\sum_{x=0}^{\infty} f(x)\cdot x(x-1)(x-2) \cdots (x-r+1)$ $\mu'_{(r)} = \left(\frac{d^r P(t)}{dt^r}\right)_{t=1}$

(*continued*)

Table 2.1: (*Continued*)

Term	Symbol	Description and notes
32. *r*th Factorial moment about the mean (discrete nonnegative variate)	$\mu_{(r)}$	$\sum_{x=0}^{\infty} f(x) \cdot (x-\mu)(x-\mu-1)\cdots(x-\mu-r+1)$

Table 2.2: General Relationships Between Moments

Moments about the origin	$\mu'_r = \sum_{i=0}^{r} \binom{r}{i} \mu_{r-i}(\mu'_1)^i,\ \mu_0 = 1$
Central moments about mean	$\mu_r = \sum_{i=0}^{r} \binom{r}{i} \mu'_{r-i}(-\mu'_1)^i,\ \mu'_0 = 1$
Hence,	$\mu_2 = \mu'_2 - \mu_1'^2$ $\mu_3 = \mu'_3 - 3\mu'_2\mu'_1 + 2\mu_1'^3$ $\mu_4 = \mu'_4 - 4\mu'_3\mu'_1 + 6\mu'_2\mu_1'^2 - 3\mu_1'^4$
Moments and cumulants	$\mu'_r = \sum_{i=1}^{r} \binom{r-1}{i-1} \mu'_{r-i}\kappa_i$

Table 2.3: Samples

Term	Symbol	Description and notes
Sample data	x_i	x_i is an observed value of a random variable.
Sample size	n	The number of observations in a sample.
Sample mean	$\bar{x}$	$\frac{1}{n}\sum_{i=1}^{n} x_i$
Sample variance (unadjusted for bias)	s^2	$\frac{1}{n}\sum_{i=1}^{n}(x_i - \bar{x})^2$
Sample variance (unbiased)	s_u^2	$\left(\frac{1}{n-1}\right)\sum_{i=1}^{n}(x_i - \bar{x})^2$

Chapter 3

General Variate Relationships

3.1 INTRODUCTION

This chapter is concerned with general relationships between variates and with the ideas and notation needed to describe them. Some definitions are given, and the relationships between variates under one-to-one transformations are developed. Location, scale, and shape parameters are then introduced, and the relationships between functions associated with variates that differ only in regard to location and scale are listed. The relationship of a general variate to the rectangular variate is derived, and finally the notation and concepts involved in dealing with variates that are related by many-to-one functions and by functionals are discussed.

Following the notation introduced in Chapter 2 we denote a general variate by X, its range by $\Re_X$, its quantile by x, and a realization or random number of X by x_X.

3.2 FUNCTION OF A VARIATE

Let ϕ be a function mapping from $\Re_X$ into a set we shall call $\Re_{\phi(X)}$.

Definition 3.2.1 (Function of a Variate). The term $\phi(X)$ is a variate such that if x_X is a random number of X then $\phi(x_X)$ is a random number of $\phi(X)$.

Thus a function of a variate is itself a variate whose value at any realization is obtained by applying the appropriate transformation to the value realized by the original variate. For example, if X is the number of heads obtained when three coins are tossed, then X^3 is the cube of the number of heads obtained. (Here, as in Chapter 2, we use the symbol X for both a variate and a random variable that is an element of that variate.)

The probabilistic relationship between X and $\phi(X)$ will depend on whether more than one number in $\Re_X$ maps into the same $\phi(x)$ in $\Re_{\phi(X)}$. That is, it is important to consider whether ϕ is or is not a one-to-one function over the range considered. This point is taken up in Section 3.3.

Statistical Distributions, Fourth Edition, by Catherine Forbes, Merran Evans, Nicholas Hastings, and Brian Peacock

A definition similar to 3.2.1 applies in the case of a function of several variates; we shall detail the case of a function of two variates. Let X, Y be variates with ranges $\Re_X, \Re_Y$ and let ψ be a functional mapping from the Cartesian product of $\Re_X$ and $\Re_Y$ into (all or part of) the real line.

Definition 3.2.2 (Function of Two Variates). The term $\psi(X, Y)$ is a variate such that if x_X and x_Y are random numbers of X and Y, respectively, then $\phi(x_X, x_Y)$ is a random number of $\psi(X, Y)$.

3.3 ONE-TO-ONE TRANSFORMATIONS AND INVERSES

Let ϕ be a function mapping from the real line into the real line.

Definition 3.3.1 (One-to-One Function). The function ϕ is one to one if there are no two numbers x_1, x_2 in the domain of ϕ such that $\phi(x_1) = \phi(x_2)$ and $x_1 \neq x_2$.

A sufficient condition for a real function to be one to one is that it be increasing in x. As an example, $\phi(x) = \exp(x)$ is a one-to-one function, but $\phi(x) = x^2$ is not (unless x is confined to all negative or all positive values, say) since $x_1 = 2$ and $x_2 = -2$ gives $\phi(x_1) = \phi(x_2) = 4$. Figures 3.1 and 3.2 illustrate this. A function that is not one to one is a *many-to-one function*. See also Section 3.8.

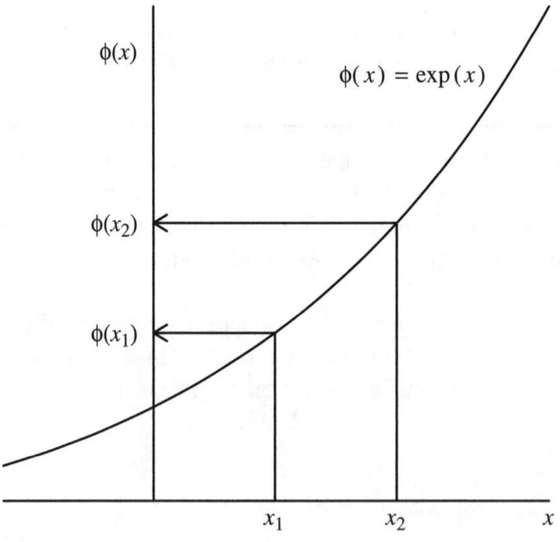

Figure 3.1. A one-to-one function.

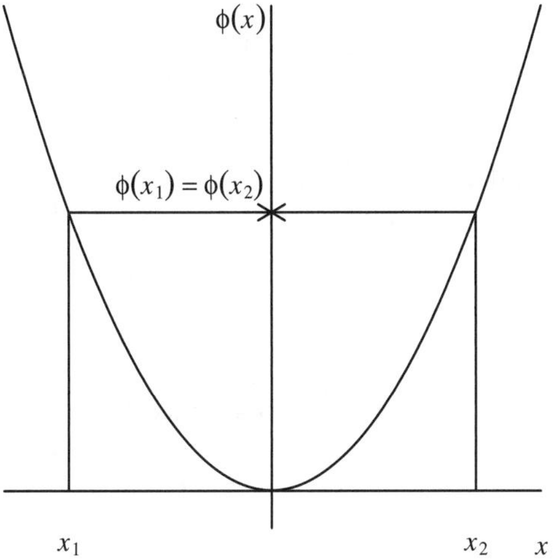

Figure 3.2. A many-to-one function.

Inverse of a One-to-One Function

The inverse of a one-to-one function ϕ is a one-to-one function ϕ^{-1}, where

$$\phi^{-1}(\phi(x)) = x, \qquad \phi(\phi^{-1}(y)) = y \tag{3.1}$$

and x and y are real numbers (Bernstein's Theorem).

3.4 VARIATE RELATIONSHIPS UNDER ONE-TO-ONE TRANSFORMATION

Probability Statements

Definitions 3.2.1 and 3.3.1 imply that if X is a variate and ϕ is an increasing one-to-one function, then $\phi(X)$ is a variate with the property

$$\Pr[X \le x] = \Pr[\phi(X) \le \phi(x)] \qquad x \in \Re_X; \phi(x) \in \Re_{\phi(X)}. \tag{3.2}$$

Distribution Function

In terms of the distribution function $F_X(x)$ for variate X at quantile x, Equation (3.2) is equivalent to the statement

$$F_X(x) = F_{\phi(X)}(\phi(x)). \tag{3.3}$$

To illustrate Equations (3.2) and (3.3) consider the experiment of tossing three coins and the random variables "number of heads," denoted by X, and "cube of the number of heads," denoted by X^3. The probability statements and distribution functions at quantiles 2 heads and 8 (heads)^3 are

$$\begin{aligned} \Pr[X \le 2] &= \Pr[X^3 \le 8] = \tfrac{7}{8} \\ F_X(2) &= F_{X^3}(8) = \tfrac{7}{8}. \end{aligned} \tag{3.4}$$

Inverse Distribution Function

The inverse distribution function (introduced in Section 2.4) for a variate X at probability level α is $G_X(\alpha)$. For a one-to-one function ϕ we now establish the relationship between the inverse distribution functions of the variates X and $\phi(X)$.

Theorem 3.4.1.

$$\phi(G_X(\alpha)) = G_{\phi(X)}(\alpha).$$

Proof Equations (2.4) and (3.3) imply that if

$$G_X(\alpha) = x \quad \text{then} \quad G_{\phi(X)}(\alpha) = \phi(x)$$

which implies that the theorem is true. □

We illustrate this theorem by extending the example of Equation (3.4). Considering the inverse distribution function, we have

$$G_X\left(\frac{7}{8}\right) = 2; \qquad G_{X^3}\left(\frac{7}{8}\right) = 8 = 2^3 = \left(G_X\left(\frac{7}{8}\right)\right)^3.$$

Equivalence of Variates

For any two variates X and Y, the statement $X \sim Y$, read "X is distributed as Y," means that the distribution functions of X and Y are identical. All other associated functions, sets, and probability statements of X and Y are therefore also identical. "Is distributed as" is an equivalent relation, so that

1. $X \sim X$.
2. $X \sim Y$ implies $Y \sim X$.
3. $X \sim Y$ and $Y \sim Z$ implies $X \sim Z$.

Inverse Function of a Variate

Theorem 3.4.2. *If X and Y are variates and ϕ is an increasing one-to-one function, then $Y \sim \phi(X)$ implies $\phi^{-1}(Y) \sim X$.*

Proof

$$Y \sim \phi(X) \quad \text{implies} \quad \Pr[Y \leq x] = \Pr[\phi(X) \leq x]$$

(by the equivalence of variates, above)

$$\begin{aligned}\Pr[Y \leq x] &= \Pr[X \leq \phi^{-1}(x)] \\ &= \Pr[\phi^{-1}(Y) \leq \phi^{-1}(x)]\end{aligned}$$

(from Equations (3.1) and (3.2)).

These last two equations together with the equivalence of variates (above) imply that Theorem 3.4b is true. □

3.5 PARAMETERS, VARIATE, AND FUNCTION NOTATION

Every variate has an associated distribution function. Some groups of variates have distribution functions that differ from one another only in the values of certain parameters. A generalized distribution function in which the parameters appear as symbols corresponds to a family of variates (not to be confused with a distribution family). Examples are the variate families of the normal, lognormal, beta, gamma, and exponential distributions. The detailed choice of the parameters that appear in a distribution function is to some extent arbitrary. However, we regard three types of parameter as "basic" in the sense that they always have a certain physical geometrical meaning. These are the location, scale, and shape parameters, the descriptions of which are as follows:

Location Parameter, a. The abscissa of a location point (usually the lower or midpoint) of the range of the variate.

Scale Parameter, b. A parameter that determines the scale of measurement of the quantile x.

Shape Parameter, c. A parameter that determines the shape (in a sense distinct from location and scale) of the distribution function (and other functions) within a family of shapes associated with a specified type of variate.

The symbols a, b, c will be used to denote location, scale, and shape parameters in general, but other symbols may be used in cases where firm conventions are established. Thus for the normal distribution the mean, μ, is a location parameter (the locating point is the midpoint of the range) and the standard deviation, σ, is a scale parameter. The normal distribution does not have a shape parameter. Some distributions (e.g., the beta) have two shape parameters, which we denote by ν and ω.

Variate and Function Notation

A variate X with parameters a, b, c is denoted in full by X: a, b, c. Some or all of the parameters may be omitted if the context permits.

The distribution function for a variate X: c is $F_X(x : c)$. If the variate name is implied by the context, we write $F(x : c)$. Similar usages apply to other functions. The inverse distribution function for a variate X: a, b, c at probability level α is denoted $G_X(\alpha : a, b, c)$.

3.6 TRANSFORMATION OF LOCATION AND SCALE

Let X: 0, 1 denote a variate with location parameter $a = 0$ and scale parameter $b = 1$. (This is often referred to as the standard variate.) A variate that differs from X: 0, 1 only in regard to location and scale is denoted X: a, b and is defined by

$$X : a, b \sim a + b(X : 0, 1). \tag{3.5}$$

The location and scale transformation function is the one-to-one function

$$\phi(x) = a + bx$$

and its inverse is

$$\phi^{-1}(x) = (x - a)/b$$

with b $\neq$ 0. (Typically b $>$ 0.)

The following equations relating to variates that differ only in relation to location and scale parameters then hold:

$$X : a, b \sim a + b(X : 0, 1)$$

(by definition)

$$X : 0, 1 \sim [(X : a, b) - a]/b$$

(by Theorem 3.4.2 and Equation (3.5)]

$$\Pr[(X : a, b) \le x] = \Pr[(X : 0, 1) \le (x - a)/b]$$

(by Equation (3.2)) (3.6)

$$F_X(x : a, b) = F_X\{[(x - a)/b] : 0, 1\}$$

(equivalent to Equation (3.6))

$$G_X(\alpha : a, b) = a + b(G_X(\alpha : 0, 1))$$

(by Theorem 3.4.1)

These and other interrelationships between functions associated with variates that differ only in regard to location and scale parameters are summarized in Table 3.1. The functions themselves are defined in Table 2.1.

3.7 TRANSFORMATION FROM THE RECTANGULAR VARIATE

The following transformation is often useful for obtaining random numbers of a variate X from random numbers of the unit rectangular variate R. The latter has distribution

Table 3.1: Relationships Between Functions for Variates that Differ Only by Location and Scale Parameters *a, b*

Variate relationship	$\boldsymbol{X}: a, b \sim a + b(\boldsymbol{X} : 0, 1)$
Probability statement	$\Pr[(\boldsymbol{X} : a, b) \leq x] = \Pr[(\boldsymbol{X} : 0, 1) \leq (x - a)/b]$
Function relationships	
Distribution function	$F(x : a, b) = F([(x - a)/b] : 0, 1)$
Probability density function	$f(x : a, b) = (1/b) f([(x - a)/b] : 0, 1)$
Inverse distribution function	$G(\alpha : a, b) = a + bG(\alpha : 0, 1)$
Survival function	$S(x : a, b) = S([(x - a)/b] : 0, 1)$
Inverse survival function	$Z(\alpha : a, b) = a + bZ(\alpha : 0, 1)$
Hazard function	$h(x : a, b) = (1/b)h([(x - a)/b] : 0, 1)$
Cumulative hazard function	$H(x : a, b) = H([(x - a)/b] : 0, 1)$
Moment generating function	$M(t : a, b) = \exp(at)M(bt : 0, 1)$
Laplace transform	$f^*(s : a, b) = \exp(-as) f^*(bs : 0, 1), a > 0$
Characteristic function	$C(t : a, b) = \exp(iat)C(bt : 0, 1)$
Cumulant function	$K(t : a, b) = iat + K(bt : 0, 1)$

function $F_{\boldsymbol{R}}(x) = x,\ 0 \leq x \leq 1$, and inverse distribution function $G_{\boldsymbol{R}}(\alpha) = \alpha, 0 \leq \alpha \leq 1$. The inverse distribution function of a general variate $\boldsymbol{X}$ is denoted $G_{\boldsymbol{X}}(\alpha)$, $\alpha \in \Re^{a}_{\boldsymbol{X}}$. Here $G_{\boldsymbol{X}}(\alpha)$ is a one-to-one function.

Theorem 3.7.1. $\boldsymbol{X} \sim G_{\boldsymbol{X}}(\boldsymbol{R})$ *for continuous variates.*

Proof

$$\Pr[\boldsymbol{R} \leq \alpha] = \alpha, 0 \leq \alpha \leq 1$$

$$\text{(property of } \boldsymbol{R}\text{)}$$

$$= \Pr[G_{\boldsymbol{X}}(\boldsymbol{R}) \leq G_{\boldsymbol{X}}(\alpha)]$$

$$\text{(by Equation (3.2))}$$

Hence, by these two equations and Equation (2.4),

$$G_{\boldsymbol{X}}(\boldsymbol{R}) \sim \boldsymbol{X}. \quad \square$$

For discrete variates, the corresponding expression is

$$\boldsymbol{X} \sim G_{\boldsymbol{X}}[f(\boldsymbol{R})], \quad \text{where} \quad f(\alpha) = \text{Min}\{p | p \geq \alpha,\ p \in \Re^{\alpha}_{\boldsymbol{X}}\}.$$

Thus every variate is related to the unit rectangular variate via its inverse distribution function, although, of course, this function will not always have a simple algebraic form.

3.8 MANY-TO-ONE TRANSFORMATIONS

In Sections 3.3 through 3.7 we considered the relationships between variates that were linked by a one-to-one function. Now we consider many-to-one functions, which are defined as follows. Let ϕ be a function mapping from the real line into the real line.

Definition 3.8.1 (Many-to-One Function). The function ϕ is many to one if there are at least two numbers x_1, x_2 in the domain of ϕ such that $\phi(x_1) = \phi(x_2)$, $x_1 \neq x_2$.

The many-to-one function $\phi(x) = x^2$ is illustrated in Figure 3.2.

In Section 3.2 we defined, for a general variate X with range $\Re_X$ and for a function ϕ, a variate $\phi(X)$ with range $\Re_{\phi(X)}$. Here $\phi(X)$ has the property that if x_X is a random number of X, then $\phi(x_X)$ is a random number of $\phi(X)$. Let r_2 be a subset of $\Re_{\phi(X)}$ and r_1 be the subset of $\Re_X$, which ϕ maps into r_2. The definition of $\phi(X)$ implies that

$$\Pr[X \in r_1] = \Pr[\phi(X) \in r_2].$$

This equation enables relationships between X and $\phi(X)$ and their associated functions to be established. If ϕ is many-to-one, the relationships will depend on the detailed form of ϕ.

EXAMPLE 3.8.1

As an example we consider the relationships between the variates X and X^2 for the case where $\Re_X$ is the real line. We know that $\phi : x \to x^2$ is a many-to-one function. In fact it is a two-to-one function in that $+x$ and $-x$ both map into x^2. Hence the probability that an as-yet unrealized random number of X^2 will be greater than x^2 will be equal to the probability that an as-yet unrealized random number of X will be either greater than $+|x|$ or less than $-|x|$.

$$\Pr[X^2 > x^2] = \Pr[X > +|x|] + \Pr[X] < -|x|]. \qquad (3.7)$$

Symmetrical Distributions

Let us now consider a variate X whose probability density function is symmetrical about the origin. We shall derive a relationship between the distribution function of the variates X and X^2 under the condition that X is symmetrical. An application of this result appears in the relationship between the F (variance ratio) and Student's t variates.

Theorem 3.8.1. *Let* **X** *be a variate whose probability density function is symmetrical about the origin.*

1. *The distribution functions $F_X(x)$ and $F_{X^2}(x^2)$ for the variates* **X** *and* $\mathbf{X}^2$ *at quantiles $x \geq 0$ and x^2, respectively, are related by*

$$F_X(x) = \tfrac{1}{2}\left[1 + F_{X^2}(x^2)\right]$$

or

$$F_{X^2}(x^2) = 2F_X(x) - 1.$$

2. *The inverse survival functions $Z_X(\frac{1}{2}\alpha)$ and $Z_{X^2}(\alpha)$ for the variates X and X^2 at probability levels $\frac{1}{2}\alpha$ and α, respectively, and with $0 < \alpha < 1$ are related by*

$$[Z_X(\tfrac{1}{2}\alpha)]^2 = Z_{X^2}(\alpha).$$

Proof

1. For a variate X with symmetrical pdf about the origin we have

$$\Pr[X > x] = \Pr[X \leq -x].$$

This and Equation (3.7) imply

$$\Pr[X^2 > x^2] = 2\Pr[X > x]. \tag{3.8}$$

Introducing the distribution function $F_X(x)$, we have, from the definition (Equation (2.1))

$$1 - F_X(x) = \Pr[X > x].$$

This and Equation (3.8) imply

$$1 - F_{X^2}(x^2) = 2[1 - F_X(x)].$$

Rearrangement of this equation gives

$$F_X(x) = \tfrac{1}{2}[1 + F_{X^2}(x^2)]. \tag{3.9}$$

2. Let $F_X(x) = \alpha$. Equation (3.9) implies

$$\tfrac{1}{2}[1 + F_{X^2}(x^2)] = \alpha$$

which can be arranged as

$$F_{X^2}(x^2) = 2\alpha - 1.$$

This and Equations (2.2) and (2.3) imply

$$G_X(\alpha) = x \quad \text{and} \quad G_{X^2}(2\alpha - 1) = x^2$$

which implies

$$[G_X(\alpha)]^2 = G_{X^2}(2\alpha - 1). \tag{3.10}$$

From the definition of the inverse survival function Z (Table 2.1, item 6), we have $G(\alpha) = Z(1 - \alpha)$. Hence from Equation (3.10)

$$[Z_X(1 - \alpha)]^2 = Z_{X^2}(2(1 - \alpha))$$

$$[Z_X(\alpha)]^2 = Z_{X^2}(2\alpha)$$

$$[Z_X(\alpha/2)]^2 = Z_{X^2}(\alpha) \qquad \square$$

Chapter 4

Multivariate Distributions

4.1 JOINT DISTRIBUTIONS

Probability statements concerning more than one random variate are called *joint* probability statements. Joint probability statements can be made about a combination of variates, all having continuous domain, all having countable domain, or some combination of continuous and countable domains.

To keep notation to a minimum, consider the case of only two variates, X and Y, and call the pair (X, Y) a *bivariate*, with each of X and Y alone referred to as *univariates*. In the general case of more than one variate (including the bivariate case), the collection of variates is more generally referred to as a *multivariate*.

Joint Range

Let $\Re_{X,Y}$ be the set of all pairs of (real number) values that the bivariate (X, Y) can take. The set $\Re_{X,Y}$ is the *joint range* of (X, Y).

Bivariate Quantile

Let the real valued pair (x, y) denote a general element of the joint range $\Re_{X,Y}$. We refer to (x, y) as a *bivariate quantile* of (X, Y).

Joint Probability Statement

Let $\Pr[X \leq x, Y \leq y]$ mean "the probability that the value realized by the univariate X is less than or equal to x and the value realized by the univariate Y is less than or equal to y." In this case, both conditions $X \leq x$ and $Y \leq y$ hold simultaneously.

Statistical Distributions, Fourth Edition, by Catherine Forbes, Merran Evans, Nicholas Hastings, and Brian Peacock

Joint Probability Domain

Let $\alpha \in [0, 1]$ and let $\Re^{\alpha}_{X,Y}$ be the set of all values (of probability) that $\Pr[X \leq x, Y \leq y]$ can take. When X and Y are jointly continuous, $\Re^{\alpha}_{X,Y}$ is the line segment $[0, 1]$. In all other cases, $\Re^{\alpha}_{X,Y}$ may be a subset of $[0, 1]$. Analogous to the univariate case, $\Re^{\alpha}_{X,Y}$ is called the *joint probability domain* of the bivariate (X, Y).

Joint Distribution Function

The *distribution function (joint df)* F associated with a bivariate (X, Y) maps from the joint range $\Re_{X,Y}$ into the joint probability domain $\Re^{\alpha}_{X,Y}$ and is such that

$$F(x, y) = \Pr[X \leq x, Y \leq y] = \alpha, \qquad \text{for } (x, y) \in \Re_{X,Y},\ \alpha \in \Re^{\alpha}_{X,Y}.$$

The function $F(x, y)$ is nondecreasing in each of x and y, and attains a value of unity when both x and y are at their respective maxima.

Joint Probability Density Function

When (X, Y) have a jointly continuous probability domain, the function $f(x, y)$ is a *bivariate probability density function (joint pdf)* if

$$F(x, y) = \int_{-\infty}^{y} \int_{-\infty}^{x} f\left(u_x, u_y\right) du_x\, du_y$$

for all (x, y) in $\Re_{X,Y}$. This relationship implies that the pdf satisfies

$$f(x, y) = \frac{\partial^2 F(x, y)}{\partial x\, \partial y}$$

and that the probability associated with an arbitrary bivariate quantile set A is equal to the area under the bivariate pdf surface A. That is,

$$\Pr[(x, y) \in A] = \iint_{(x,y)\in A} f(x, y)\, dx\, dy.$$

Joint Probability Function

A discrete bivariate (X, Y) takes discrete values (x, y) with finite probabilities $f(x, y)$. In this case $f(x, y)$ is the *bivariate probability function (joint pf)*.

Table 4.1: Joint Probability Function for Discrete Bivariate Example

$\Pr[X = 0, Y = 0] = 1/8$	$\Pr[X = 0, Y = 1] = 1/8$
$\Pr[X = 1, Y = 0] = 3/8$	$\Pr[X = 1, Y = 1] = 1/8$
$\Pr[X = 2, Y = 0] = 3/16$	$\Pr[X = 2, Y = 1] = 1/16$

EXAMPLE 4.1.1. ***Discrete Bivariate***

Consider the discrete bivariate (X, Y) with joint distribution function given by

$$F(x, y) = \begin{cases} 0 & x < 0, y < 0 \\ 1/8 & 0 \leq x < 1, 0 \leq y < 1 \\ 1/4 & 0 \leq x < 1, y \geq 1 \\ 1/2 & 1 \leq x < 2, 0 \leq y < 1. \\ 3/4 & 1 \leq x < 2, y \geq 1 \\ 11/16 & x \geq 2, 0 \leq y < 1 \\ 1 & x \geq 2, y \geq 1. \end{cases}$$

As the joint distribution function changes only when x reaches zero, one or two, and when y reaches zero or one, the joint range is $\Re_{X,Y} = \{(0, 0), (1, 0), (2, 0), (0, 1), (1, 1), (1, 2)\}$. The resulting joint pf is given in Table 4.1. In this case the joint probability domain is $\Re^{\alpha}_{X,Y} = \left\{\frac{1}{8}, \frac{1}{4}, \frac{1}{2}, \frac{3}{4}, \frac{11}{16}, 1\right\}$.

4.2 MARGINAL DISTRIBUTIONS

Probabilities associated with a univariate element of a bivariate without regard to the value of the other univariate element within the same bivariate arise from the *marginal distribution* of that univariate. Corresponding to such a marginal distribution are the (univariate) range, quantile, and probability domain associated with the univariate element, with each one consistent with the corresponding bivariate entity.

Marginal Probability Density Function and Marginal Probability Function

In the case of continuous bivariate (X, Y) having joint pdf $f(x, y)$, the *marginal probability density function* (*marginal pdf*) of the univariate X is found for each quantile x by integrating the joint pdf over all possible bivariate quantiles $(x, y) \in \Re_{X,Y}$ having fixed value for x,

$$f(x) = \int_{-\infty}^{\infty} f(x, y)\, dy.$$

Table 4.2: Marginal Probability Function for X from Discrete Bivariate Example

$\Pr(X=0)$	$=$	$\Pr(X=0, Y=0) + \Pr(X=0, Y=1)$	$=$	$1/4$
$\Pr(X=1)$	$=$	$\Pr(X=1, Y=0) + \Pr(X=1, Y=1)$	$=$	$1/2$
$\Pr(X=2)$	$=$	$\Pr(X=2, Y=0) + \Pr(X=2, Y=1)$	$=$	$1/4$

Table 4.3: Marginal Probability Function for Y from Discrete Bivariate Example

$\Pr[Y=0]$	$=$	$\Pr[X=0, Y=0] + \Pr[X=1, Y=0] + \Pr[X=2, Y=0]$
	$=$	$11/16$
$\Pr[Y=1]$	$=$	$\Pr[X=0, Y=1] + \Pr[X=1, Y=1] + \Pr[X=2, Y=1]$
	$=$	$5/16$

Similarly, the marginal pdf of the univariate Y is

$$f(y) = \int_{-\infty}^{\infty} f(x, y)\, dx.$$

In the case of discrete variates X and Y, the *marginal probability function* (*marginal pf*) of X is found for each quantile x by summing the joint pf associated with the bivariate quantiles $(x, y) \in \Re_{X,Y}$ having fixed x

$$f(x) = \sum_{y \in \Re_y} f(x, y).$$

The *marginal pf* of Y is

$$f(y) = \sum_{x \in \Re_x} f(x, y).$$

EXAMPLE 4.2.1. ***Discrete Bivariate Continued***

Continuing Example 4.1.1, the marginal pf of X, evaluated at each value of $X = x$, is given in Table 4.2. The corresponding marginal domain for X is $\Re_X = \{0, 1, 2\}$ with marginal probability domain $\Re_X^\alpha = \{1/4, 3/4, 1\}$. Similarly, the marginal pf of Y is given in Table 4.3. The corresponding marginal domain for Y is $\Re_Y = \{0, 1\}$ with marginal probability domain $\Re_Y^\alpha = \{11/16, 1\}$.

4.3 INDEPENDENCE

Two univariates are said to be *independent* if the marginal probabilities associated with the outcomes of one variate are not influenced by the observed value of the other variate, and vice versa. This implies that the joint pdf for a continuous bivariate with independent continuous univariate elements may be decomposed into the product of the marginal pdfs. Similarly, the joint pf for a discrete bivariate with independent discrete univariate elements may be decomposed into the product of the marginal pfs. That is,

$$f(x, y) = f(x) f(y) \quad \text{for all} \quad (x, y) \in \Re_{X,Y}$$

4.4 CONDITIONAL DISTRIBUTIONS

When two univariates are not independent, they are said to be *dependent*. In this case, interest often focuses on computing probabilities associated with one variate given information about the realized value of the other variate. Such probabilities are called *conditional probabilities*. For example, the "conditional probability that $\boldsymbol{X}$ is less than x, *given it is known that* $\boldsymbol{Y}$ *is less than* y" is

$$\Pr[\boldsymbol{X} \leq x | \boldsymbol{Y} \leq y] = \frac{\Pr[\boldsymbol{X} \leq x, \boldsymbol{Y} \leq y]}{\Pr[\boldsymbol{Y} \leq y]}$$

provided the denominator $\Pr[\boldsymbol{Y} \leq y]$ is nonzero.

Conditional Probability Function and Conditional Probability Density Function

When conditioning on a single observed value of one univariate component $\boldsymbol{Y} = y$ of a discrete bivariate $(\boldsymbol{X}, \boldsymbol{Y})$, the conditional pf of $\boldsymbol{X}$, given $\boldsymbol{Y} = y$, and denoted by $f(x|y)$, is found by taking the ratio of the joint pf $f(x, y)$ to the marginal pf of $\boldsymbol{Y}$, $f(y)$, for fixed $\boldsymbol{Y} = y$

$$f(x|y) = \frac{\Pr[X = x, Y = y]}{\Pr[Y = y]} = \frac{f(x, y)}{f(y)},$$

for all $(x, y) \in \Re_{X,Y}$. Similarly,

$$f(y|x) = \frac{\Pr[X = x, Y = y]}{\Pr[X = x]} \quad \text{for all} \quad (x, y) \in \Re_{X,Y}.$$

For continuous random variates, the *conditional probability density function* (*conditional pdf*) for $\boldsymbol{X}$ *given* $\boldsymbol{Y} = y$ and denoted by $f(x|y)$ may be constructed using the ratio of the joint pdf over the marginal pdf of $\boldsymbol{Y}$, for fixed $\boldsymbol{Y} = y$

$$f(x|y) = \frac{f(x, y)}{f(y)} \quad \text{for all} \ (x, y) \in \Re_{X,Y},$$

for all $(x, y) \in \Re_{X,Y}$. Similarly,

$$f(y|x) = \frac{f(x, y)}{f(x)} \quad \text{for all} \quad (x, y) \in \Re_{X,Y}.$$

Composition

Joint, marginal, and conditional probabilities satisfy the property of *composition*, whereby joint probabilities may be determined by the product of marginal probabilities and the appropriate conditional probability

$$\Pr[X \le x, Y \le y] = \Pr[X \le x | Y \le y]\Pr[Y \le y].$$

The composition property implies joint, marginal, and conditional pdfs (or pfs) satisfy

$$f(x, y) = f(x|y)\, f(y) = f(y|x)\, f(x) \text{ for all } (x, y) \in \Re_{X,Y}. \tag{4.1}$$

Equation (4.1) holds whether the variables are dependent or independent. However, if X and Y are independent, then $f(y|x) = f(y)$ and $f(x|y) = f(x)$, and hence

$$f(x, y) = f(x)f(y).$$

EXAMPLE 4.4.1. ***Discrete Bivariate Continued***

Again continuing Example 4.1.1 the conditional pf of X, given $Y = 0$, is found by considering the individual joint probabilities in Table 4.1 and dividing by the marginal probability that $Y = 0$ from Table 4.3. The resulting conditional probabilities are

$$\Pr[X = 0|Y = 0] = \frac{\Pr[X = 0, Y = 0]}{\Pr[Y = 0]} = \frac{1/8}{11/16} = 2/11$$

$$\Pr[X = 1|Y = 0] = \frac{\Pr[X = 1, Y = 0]}{\Pr[Y = 0]} = \frac{3/8}{11/16} = 6/11$$

$$\Pr[X = 2|Y = 0] = \frac{\Pr[X = 2, Y = 0]}{\Pr[Y = 0]} = \frac{3/16}{11/16} = 3/11.$$

These conditional probabilities associated with the univariate X given $Y = 0$ are not the same as those obtain by conditioning on a value of $Y = 1$, since

$$\Pr[X = 0|Y = 1] = \frac{\Pr[X = 0, Y = 1]}{\Pr[Y = 1]} = \frac{1/8}{5/16} = 2/5$$

$$\Pr[X = 1|Y = 1] = \frac{\Pr[X = 1, Y = 1]}{\Pr[Y = 1]} = \frac{1/8}{5/16} = 2/5$$

$$\Pr[X = 2|Y = 1] = \frac{\Pr[X = 2, Y = 1]}{\Pr[Y = 1]} = \frac{1/16}{5/16} = 1/5.$$

The conditional probabilities for Y given $X = x$ are similarly found, with

$$\Pr[Y = 0|X = 0] = 1/2$$

$$\Pr[Y = 1|X = 0] = 1/2$$

$$\Pr[Y = 0|X = 1] = 3/4$$

$$\Pr[Y = 1|X = 1] = 1/4$$

and

$$\Pr[Y = 0|X = 2] = 3/4$$
$$\Pr[Y = 1|X = 2] = 1/4.$$

Note, however, that composition may be used to reconstruct the joint probabilities for (X, Y), for example, with

$$\Pr[X = 0, Y = 1] = \Pr[X = 0|Y = 1]\Pr[Y = 1] = \frac{2}{5} \times \frac{5}{16} = \frac{1}{8},$$

concurring with $\Pr[X = 0, Y = 1]$ in Table 4.1.

4.5 BAYES' THEOREM

A fundamental result concerning conditional probabilities is known as Bayes' theorem, providing the rule for reversing the roles of variates in a conditional probability. Bayes' theorem states

$$f(y|x) = \frac{f(x|y)f(y)}{f(x)}. \tag{4.2}$$

For a continuous bivariate, the terms denoted by $f(\cdot)$ in Equation (4.2) relate to marginal and conditional pdfs, whereas for a discrete bivariate these terms relate to marginal and conditional pfs.

EXAMPLE 4.5.1. ***Discrete Bivariate Continued***

Using Bayes' theorem in Equation (4.2) relates the conditional probability $\Pr[X = x|Y = y]$ to the conditional probability $\Pr[Y = y|X = x]$, with

$$\Pr[X = x|Y = y] = \frac{\Pr[Y = y|X = x]\Pr[X = x]}{\Pr[Y = y]}.$$

So, for example,

$$\Pr[X = 2|Y = 0] = \frac{\Pr[Y = 0|X = 2]\Pr[X = 2]}{\Pr[Y = 0]}$$
$$= \frac{(3/4)(1/4)}{11/16} = 3/11.$$

as before.

4.6 FUNCTIONS OF A MULTIVARIATE

If (X, Y) is a bivariate with range $\Re_{X,Y}$ and ψ is a functional mapping from $\Re_{X,Y}$ into the real line, then $\psi(X, Y)$ is a variate such that if x_X and x_Y are random numbers of X and Y, respectively, then $\psi(x_X, x_Y)$ is a random number of $\psi(X, Y)$.

The relationships between the associated functions of X and Y on the one hand and of $\psi(X, Y)$ on the other are not generally straightforward and must be derived by analysis of the joint distribution of the variates in question. One important general result applies when X and Y are independent and where the function is a summation, say, $Z = X + Y$. In this case practical results may often be obtained by using a property of the characteristic function $C(t)$, namely, $C_{X+Y}(t) = C_X(t)C_Y(t)$; that is, the characteristic function of the sum of two independent variates is the product of the characteristic functions of the individual variates.

We are often interested in the sum (or other functions) of two or more variates that are independently and identically distributed. Thus consider the case $Z \sim X + Y$, where $X \sim Y$. In this case we write

$$Z \sim X_1 + X_2.$$

Note that $X_1 + X_2$ is not the same as $2X_1$, even though $X_1 \sim X_2$. The term $X_1 + X_2$ is a variate for which a random number can be obtained by choosing a random number of X_1 and then another independent random number of X_2 and then adding the two. The term $2X_1$ is a variate for which a random number can be obtained by choosing a single random number of X_1 and multiplying it by two.

If there are n such independent variates of the form $(X : a, b)$ to be summed,

$$Z \sim \sum_{i=1}^{n} (X : a, b)_i.$$

where the subscript i identifies the ith variate, for $i = 1, 2, \ldots n$. In this case the distribution of each X variate is governed by the same set of parameters, (a, b).

When the variates to be summed differ in their parameters, we write

$$Z \sim \sum_{i=1}^{n} (X : a_i, b_i).$$

Chapter 5

Stochastic Modeling

5.1 INTRODUCTION

In this chapter we review the construction of a joint probability model for a sequence of random variates, suitable for the description of observed data. Choosing a stochastic representation for observed data is arguably the most critical component of any statistical exercise. In many cases the choice will be dictated by convention and convenience, however, the implications of an incorrect model structure may be severe.

Before reviewing a few general approaches to specifying a stochastic model, it is worth pausing to consider one's objective in modeling data. For example, is the main goal to achieve a parsimonious description of the observed empirical distribution of the data? Is the prediction of future observations of interest? Alternatively, is there interest in testing whether the data are consistent with certain theoretical hypotheses? There may in fact be multiple objectives that one pursues when modeling data, but to some degree the modeling method employed will be strongly influenced by the main objective.

5.2 INDEPENDENT VARIATES

If it is reasonable to assume that variates are independent of one another and that the marginal distribution for each variate is the same, then the variates are said to be *independent and identically distributed (i.i.d.)*. When variates $X_1, X_2, \ldots, X_n$ are *i.i.d.* the joint probability density function of the multivariate $(X_1, X_2, \ldots, X_n)$ is given by the product of the marginal univariate pdfs

$$f(\boldsymbol{x}) = \prod_{i=1}^{n} f(x_i)$$

where $\boldsymbol{x} = (x_1, x_2, \ldots, x_n)$ denotes the multivariate quantile of $\boldsymbol{X}$. Variates that are not independent are said to be *dependent*. The modeling of dependent variates is discussed in Section 5.6.

Statistical Distributions, Fourth Edition, by Catherine Forbes, Merran Evans, Nicholas Hastings, and Brian Peacock

The marginal distribution function of *i.i.d.* variates each having distribution function F, will be also be F. In this case, the empirical distribution function $F_E(x)$ (Chapter 14) will serve as a useful guide to the selection of the form of F. Then one can choose a joint distribution for the observed data by simply trying to match the characteristics of the empirical distribution function to one of the many functional forms available in the remaining chapters of this book.

However, recent developments in the statistical distributions literature offer alternative methods for constructing more flexible distributional forms. We review three particular approaches here, namely

1. Mixture distributions,
2. Skew-symmetric distributions, and
3. Distributions characterized by conditional skewness.

Each of these approaches yield new distributions that are built from some of the existing distributions in Chapters 8 through 47 such that particular features of the standard distributions are relaxed. All three of the above approaches have the potential to generate distributions with a range of skewness and kurtosis characteristics; the mixture distribution approach is particularly useful for generating multi-modal distributions.

5.3 MIXTURE DISTRIBUTIONS

A mixture distribution has a distribution function with a representation as a convex combination of other specific probability distribution functions. A mixture may be comprised of a finite number of base elements, where usually a relatively small number of individual distributions are combined together, or an infinite number of base elements. Often an individual base distribution is thought of as representing a unique subpopulation within the larger (sampled) population. In both the finite and infinite case, the probability of an outcome may be thought of as a weighted average of the conditional probabilities of that outcome given each base distribution, where the relevant mixture weight describes the relative likelihood of a draw from that distribution being obtained.

Finite Mixture

A *finite mixture of two distributions* having cdfs $F_1(x)$ and $F_2(x)$, respectively, has cdf

$$F(x) = \eta F_1(x) + (1 - \eta)F_2(x),$$

as long as $0 < \eta < 1$. Extending this notion to a finite mixture of K distributions (sometimes referred to as a *finite* K*−mixture*) involves using a convex combination of distinct distribution functions $F_i(x)$, for $i = 1, 2, \ldots, K$, resulting in the finite

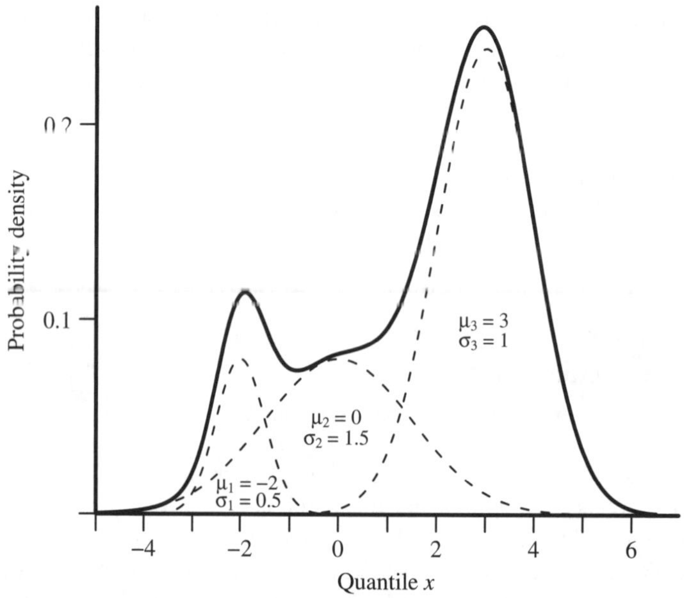

Figure 5.1. Probability density function for a 3-mixture of normal variates. Mixture weights are $\eta_1 = 0.1$, $\eta_2 = 0.3$, and $\eta_3 = 0.6$. The components are $N : \mu_i, \sigma_i$, for $i = 1, 2, 3$.

mixture cdf

$$F(x) = \sum_{i=1}^{K} \eta_i F_i(x).$$

As the combination is convex, each of the mixture weights $\eta_1, \eta_2, \ldots, \eta_K$ are between zero and one, and sum to unity.

Due to their ability to combine very different distributional structures, finite mixture distributions are well suited to cater for a large range of empirical distributions in practice. However, finite mixture models are often over-parameterized, leading to identification issues. See Frühwirth-Schnatter (2006) for a technical overview of finite mixture models and their properties, along with methodological issues concerning statistical inference and computation.

The solid curve in Figure 5.1 displays the pdf of a mixture of three normal variates. (See Chapter 33.) In this case, $\eta_1 = 0.1$, $\eta_2 = 0.3$, and $\eta_3 = 0.6$. The dashed curves indicate the three individual base components contained in the mixture. Similarly, Figure 5.2 displays a mixture of $\boldsymbol{\gamma} : b_i, c_i$ variates, for $i = 1, 2, 3$.

Moments

When the K individual distributions in the finite mixture have finite rth moment about the origin, denoted $\mu'_{r,i}$ for $i = 1, 2, \ldots, K$ respectively, then the moment about the

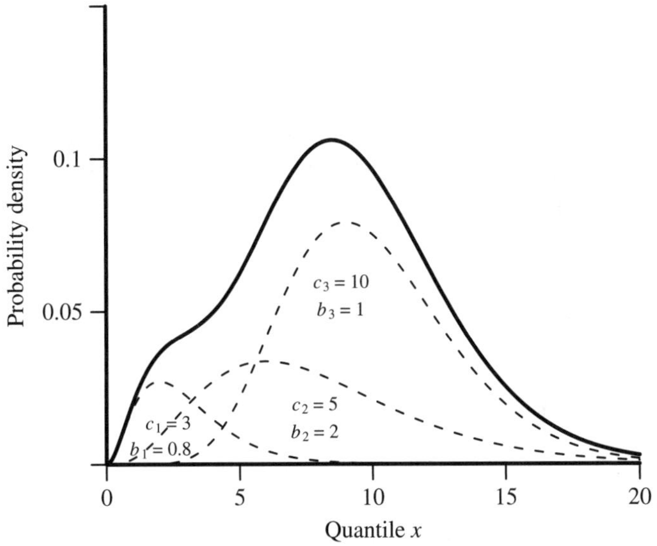

Figure 5.2. Probability density function for a 3-mixture of gamma variates. Mixture weights are $\eta_1 = 0.1$, $\eta_2 = 0.3$, and $\eta_3 = 0.6$. The components are $\boldsymbol{\gamma} : b_i, c_i$, for $i = 1, 2, 3$.

origin for the mixture distribution, μ'_r, satisfies

$$\mu'_r = \sum_{i=1}^{K} \eta_i \mu'_{r,i}.$$

Higher order moments about the mean for a mixture distribution may be constructed from the moments about the origin in the usual way, as shown in Table 2.2.

Simulation

Simulation of a variate from a finite K-mixture distribution is undertaken in two steps. First a multivariate $\boldsymbol{M} : 1, \eta_1, \ldots, \eta_K$ mixture indicator variate is drawn from the multinomial distribution with K probabilities equal to the mixture weights. Then, given the drawn mixture indicator value, k say, the variate $\boldsymbol{X}$ is drawn from the kth component distribution. The mixture indicator value k used to generate the $\boldsymbol{X} = x$ is then discarded.

Infinite Mixture of Distributions

An extension of the notation for finite mixtures that allows for a possibly infinite mixture of distributions may be constructed through marginalization of a bivariate distribution. Recall that for a continuous bivariate $(\boldsymbol{X}, \boldsymbol{Y})$ having joint pdf $f(x, y)$, the marginal density of X is given by

$$f(x) = \int_{-\infty}^{\infty} f(x|y)f(y)dy.$$

Due to the properties of the marginal pdf of Y, $f(y)$ is always (almost surely) non negative with a unit integral. Hence, the marginal distribution of X is seen as a mixture of the conditional distributions of X given $Y = y$ with the mixture weights determined by the marginal density function of Y. Note that the mixture distribution function also satisfies

$$F(x) = \int_{-\infty}^{\infty} F(x|y)f(y)dy.$$

Moments

When each of the individual conditional distributions in the infinite mixture have a finite rth moment about the origin,

$$\mu'_{r,y} = \int_{-\infty}^{\infty} x^r f(x|y)dx$$

and the unconditional moment about the origin for the mixture distribution, μ'_r, exists and is finite, then μ'_r satisfies the iterated conditional moment relationship

$$\begin{aligned}\mu'_r &= \int_{-\infty}^{\infty}\int_{-\infty}^{\infty} x^r f(x|y)\, f(y)\, dx\, dy \\ &= \int_{-\infty}^{\infty} \mu'_{r,y}\, f(y)\, dy\end{aligned}$$

resulting in the unconditional mean

$$\mu_1 = \int_{-\infty}^{\infty} \mu'_{1,y}\, f(y)\, dy \tag{5.1}$$

and unconditional variance

$$\mu_2 = \int_{-\infty}^{\infty} \mu'_{2,y}\, f(y)\, dy - (\mu_1)^2. \tag{5.2}$$

As the case with finite mixtures, higher order central moments for the infinite mixture distributions may be constructed from the moments about the origin in the usual way, as shown in Table 2.2.

EXAMPLE 5.3.1. ***Student's t***

In some cases, particularly with infinite mixtures, a known distribution can be shown to be equivalent to a particular mixture distribution. An example of such a distribution is the Student's t distribution. If conditioned on the value $Y = y$, the variate X is a normal variate $N : \mu = 0, \sigma = (cy/\lambda)^{1/2}$ having mean zero and variance cy/λ, and if marginally Y is an inverted gamma variate $\boldsymbol{Y} \sim 1/(\boldsymbol{\gamma} : \lambda = 1/b, c)$ having mean $\lambda/(c-1)$ and variance $\lambda^2/(c-1)^2(c-2)$ then the marginal distribution of $\boldsymbol{X}$ alone is that of $\boldsymbol{t} : \nu = 2c$. To prove this, note from Chapter 33 the pdf of the normal variate $\boldsymbol{X} \sim \boldsymbol{N} : o, (cy/\lambda)^{1/2}$ is

$$f(x|y) = \frac{\lambda^{1/2}}{(2\pi c y)^{1/2}} \exp\left(-\frac{\lambda x^2}{2cy}\right)$$

and from Chapter 22 the pdf of the inverted gamma variate $\boldsymbol{Y} \sim 1/\boldsymbol{\gamma} : 1/b, c$ is

$$f(y) = \frac{\exp\left(-\frac{\lambda}{y}\right)\lambda^c}{y^{c+1}\Gamma(c)}$$

and hence the marginal pdf of $\boldsymbol{X}$ is found to be that of a $\boldsymbol{t} : \nu = 2c$ variate, since

$$\begin{aligned}
f(x) &= \int_0^\infty f(x|y) f(y)\, dy \\
&= \int_0^\infty \frac{\lambda^{c+\frac{1}{2}}}{(2\pi c)^{\frac{1}{2}}\, y^{c+\frac{3}{2}}\, \Gamma(c)} \exp\left(-\frac{\lambda}{y}\left[1+\frac{x^2}{2c}\right]\right) dy \\
&= \frac{\lambda^{c+\frac{1}{2}}\,\Gamma\left(c+\frac{1}{2}\right)}{(2\pi c)^{\frac{1}{2}}\,\Gamma(c)\left(\lambda\left[1+\frac{x^2}{2c}\right]\right)^{c+\frac{1}{2}}} \times \int_0^\infty \frac{\exp\left(-\frac{\lambda}{y}\left[1+\frac{x^2}{2c}\right]\right)\left(\lambda\left[1+\frac{x^2}{2c}\right]\right)^{c+\frac{3}{2}}}{y^{c+\frac{3}{2}}\,\Gamma\left(c+\frac{1}{2}\right)} dy \\
&= \frac{\Gamma\left(\frac{(\nu+1)}{2}\right)}{(\nu\pi)^{\frac{1}{2}}\,\Gamma\left(\frac{\nu}{2}\right)\left[1+\frac{x^2}{\nu}\right]^{\frac{(\nu+1)}{2}}}
\end{aligned}$$

where $\nu = 2c$. See Chapter 43.

Moments

Note that the mean of a Student's t variate, $t : 2c$ indeed satisfies the relation in Equation (5.1), with unconditional mean $\mu_1 = 0$ since $\mu'_{1,y} = 0$, and the variance satisfies the relation in Equation (5.2), since $\mu'_{2,y} = c\,y/\lambda$ and hence

$$\begin{aligned}
\mu_2 &= \int_0^\infty \frac{c\,y}{\lambda} f(y)\, dy \;-\; 0 \\
&= \frac{c}{\lambda}\int_0^\infty y \frac{\exp\left(-\frac{\lambda}{y}\right)\lambda^c}{y^{c+1}\Gamma(c)} dy \\
&= \frac{c}{\lambda}\left[\frac{\lambda}{c-1}\right] = \frac{c}{c-1}.
\end{aligned}$$

Simulation

Simulation of a variate from an infinite mixture distribution is undertaken in two steps. First the variate $\boldsymbol{Y}$ is drawn from its marginal distribution $F(y)$. Then, given this drawn value of $Y = y_Y$ the variate X is drawn from its conditional distribution. The draw of $Y = y_Y$ is then discarded.

5.4 SKEW SYMMETRIC DISTRIBUTIONS

Azzalini (1985) proposed a class of density functions that induce skewness into the kernel of a normal density function by multiplying the standard normal pdf evaluated at x by twice the normal distribution function evaluated at λx, a scaled value of x. The resulting distribution, called the *skew-normal* distribution, can exhibit either positive or negative skewness, depending on the value of a new real-valued parameter, λ. The standard normal distribution is a special case of the skew-normal distribution corresponding to a value of $\lambda = 0$.

This method of producing skewed distributions has been extended to non-normal but symmetric kernels by Azzalini and others, and it has since been recognized that O'Hagan and Leonard (1976) used a similar approach in the context of setting a prior distribution for a Bayesian analysis. The approach reviewed here follows Nadarajah and Kotz (2003), where a density is constructed from two arbitrary absolutely continuous independent random variates, $\boldsymbol{X_1}$ and $\boldsymbol{X_2}$, both having distributions that are symmetric about zero. If $f_{X1}(x)$ denotes the pdf of $\boldsymbol{X_1}$ and $F_{X_2}(x)$ denotes the cdf of $\boldsymbol{X_2}$, then for any real-valued scalar λ, the function

$$f_\lambda(x) = 2 f_{X_1}(x) F_{X_2}(\lambda x)$$

is a valid pdf for a random variate $\boldsymbol{X}$. Various forms of F_{X_2} have been investigated in detail, with the moments and in some cases the characteristic function having been derived.

Simulation

The simulation of $\boldsymbol{X}$ is facilitated by two different representations. For the first representation, let the variate $\boldsymbol{X} = \boldsymbol{S}\,\boldsymbol{X_1}$ where

$$\boldsymbol{S} = \begin{cases} +1 \text{ with probability } F_{X_2}(\lambda x_1) \\ -1 \text{ with probability } 1 - F_{X_2}(\lambda x_1). \end{cases}$$

Then $\boldsymbol{X}$ may be simulated by first drawing $\boldsymbol{X_1}$ directly from the distribution having pdf $f_{X_1}(x_1)$, and then conditional upon $\boldsymbol{X_1} = x_1$, generate $\boldsymbol{S} = 2(\boldsymbol{B} : 1, F_{X_2}(\lambda x_1)) - 1$, where $(\boldsymbol{B} : 1, p)$ denotes a standard Bernoulli variate.

Alternatively, the second representation is that $\boldsymbol{X} = \boldsymbol{S}|\boldsymbol{X_1}|$ where

$$\boldsymbol{S} = \begin{cases} +1 \text{ with probability } F_{X_2}(\lambda |x_1|) \\ -1 \text{ with probability } 1 - F_{X_2}(\lambda |x_1|). \end{cases}$$

Hence $\boldsymbol{X}$ may be simulated by again using a draw of $\boldsymbol{X_1}$ from the distribution having pdf $f(x_1)$, but then conditional upon $\boldsymbol{X_1} = x_1$, generate $\boldsymbol{S} = 2(\boldsymbol{B} : 1, F_{X_2}(\lambda |x_1|)) - 1$.

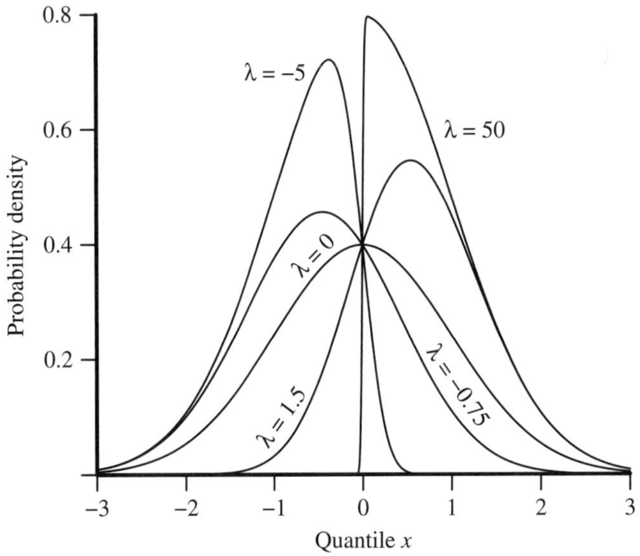

Figure 5.3. Probability density function for the skew-normal variate $SN : \lambda$.

EXAMPLE 5.4.1. ***Skew-Normal***

Azzalini's (1985) initial class of distributions take both X_1 and X_2 to be $N : 0, 1$ variates. Then

$$f_\lambda(x) = 2\,\frac{\exp(-x^2/2)}{(2\pi)^{1/2}} \int_{-\infty}^{x} \frac{\exp(-(\lambda\, u)^2/2)}{(2\pi)^{1/2}}\, du \tag{5.3}$$

Figure 5.3 displays the pdf of the skew-normal X for a selection of λ values, while Figure 5.4 displays the corresponding cdfs. Note when $\lambda = 0$ the distribution reverts to that of a $N : 0, 1$ variate, but with $\lambda < 0$ the skew-normal distribution is negatively skewed and when $\lambda > 0$ the distribution is positively skewed.

The skew-normal distribution for fixed λ is unimodal, in particular with the natural logarithm of the pdf $f_\lambda(x)$ in Equation (5.3) being a concave function of x. In addition, as $\lambda \to \infty$, the distribution of the variate X tends to the half-normal distribution.

5.5 DISTRIBUTIONS CHARACTERIZED BY CONDITIONAL SKEWNESS

Fernández and Steel (1998) suggest an alternative way to modify a symmetric density such that the result is a skewed density. Their approach relies upon taking a univariate symmetric about zero and unimodal density function $f(x)$ and a positive real-valued scalar, γ. They then define the new density function conditional on the

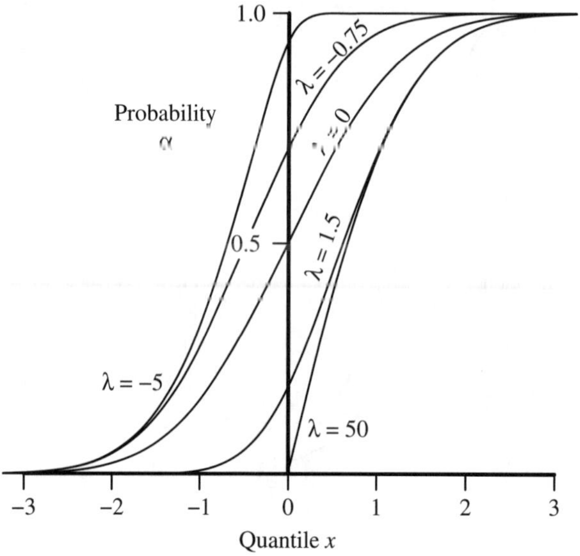

Figure 5.4. Distribution function for the skew-normal variate $SN : \lambda$.

scalar variable

$$f(x|\gamma) = \frac{2}{\gamma + \frac{1}{\gamma}} \left\{ f\left(\frac{x}{\gamma}\right) I_{[0,\infty)}(x) + f(\gamma x)\, I_{(-\infty,0)}(x) \right\}.$$

Here the parameter $\gamma > 0$ introduces skewness into the symmetric f by scaling over the entire range of x, albeit with differing impact for $x > 0$ than for $x \leq 0$. Under this framework, although the density retains its single mode at $x = 0$, the distribution remains symmetric only when $\gamma = 1$, with positive skewness associated for $\gamma > 1$ and negative skewness associated with $\gamma < 1$. The pdf for the conditional skewness model is shown for γ equal to 0.5, 1, and 4 in Figure 5.5, whereas the corresponding distribution functions are shown in Figure 5.6.

An advantage of the approach is that it leads to closed form solutions for the moments of the constructed skewed distributions in terms of the moments of the underlying symmetric distribution. In addition, the pdf of the skewed distribution retains the differentiability properties of the original pdf from the underlying distribution. The conditioning framework is especially appealing in a Bayesian setting, where models are typically constructed using composition.

Moments

Given a value of γ, the rth moment about the origin, $\mu'_{r,\gamma}$ satisfies

$$\mu'_{r,\gamma} = \mu'_r \frac{\gamma^{r+1} + \frac{(-1)^r}{\gamma^{r+1}}}{\gamma + \frac{1}{\gamma}} \tag{5.4}$$

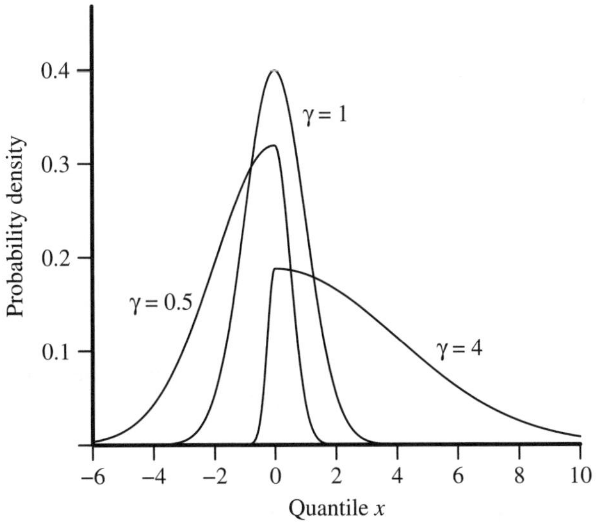

Figure 5.5. Probability function for the conditional skewness model of Fernández and Steel (1998).

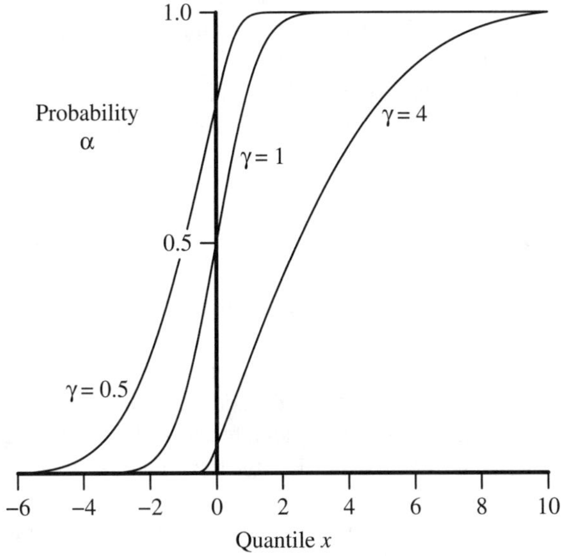

Figure 5.6. Distribution function for the conditional skewness model of Fernández and Steel (1998).

where μ'_r denotes the rth moment about the origin for the underlying symmetric distribution having pdf f. Note that the rth moment about the origin is real-valued only for integer-valued r.

From Equation (5.4) we have

$$\mu_{1,\gamma} = \mu_1' \frac{\gamma^2 - \frac{1}{\gamma^2}}{\gamma + \frac{1}{\gamma}} = \mu_1' \frac{\gamma^4 - 1}{\gamma^3 + \gamma}$$

and

$$\mu_{2,\gamma} = \mu_2' \frac{\gamma^6 + 1}{\gamma^4 + \gamma^2} - \mu^2_{1,\gamma}.$$

Higher order central moments may be constructed from the moments about the origin in the usual way shown in Table 2.2. In particular, the coefficient of skewness, $\eta_3 = \mu_3/\mu_2^{3/2}$, is

$$\eta_3 = \left(\gamma - \frac{1}{\gamma}\right) \frac{\left(\mu_3' + 2\mu_1'^3 - 3\mu_1'\mu_2'\right)\left(\gamma^2 + \frac{1}{\gamma^2}\right) + 3\mu_1'\mu_2' - 4\mu_1'^3}{\left\{\left(\mu_2' - \mu_1'^2\right)\left(\gamma^2 + \frac{1}{\gamma^2}\right) + 2\mu_1'^2 - \mu_2'\right\}^{3/2}}.$$

Due to the relative similarity of γ and $1/\gamma$ in Equation (5.4), it can be shown that when r is even, the rth moment about the origin under a value γ is identical to the rth moment about the origin under a value $1/\gamma$. However, when r is odd, the absolute value of the rth moment about the origin under a value γ is the same as the absolute value of the rth moment about the origin under a value $1/\gamma$, but one will be positive and the other will be negative.

5.6 DEPENDENT VARIATES

When observations are not independent, even when each of the individual variates $X_i, i = 1, 2, \ldots, n$ have common marginal distribution F, the empirical distribution function $F_E(x)$ need not directly reflect the form of this marginal distribution. Hence, a single univariate distribution is in general not sufficient to describe the properties of the multivariate $X = (X_1, X_2, \ldots, X_n)$. Instead, we tend to think about bringing independent variates into regression-type models that describe the relationship of the variates, one to another.

When modeling dependent variates, it is often useful to think about describing the joint density of the multivariate using conditioning. By identifying conditioning variates, one can sometimes build up from specific univariate distributions to the desired multivariate distribution via conditioning relationships. The specification of the conditioning relationships is therefore considered part of the modeling process. Many general classes of models have been developed using this approach. However, it is beyond the scope of this book to explore any but the simplest of model structures. Hence, to illustrate how to use statistical distributions within the context of modeling dependent data, we demonstrate the ideas through a relatively simple example.

EXAMPLE 5.6.1. ***Autoregressive Order One Model***

Consider the multivariate $\boldsymbol{X} = (\boldsymbol{X}_1, \boldsymbol{X}_2, \ldots, \boldsymbol{X}_n)$ and assume that there exists a natural ordering of the variates according to $t = 1, 2, \ldots, n$. The conditioning relationship given by

$$x_t = \phi\, x_{t-1} + \epsilon_t, \quad \text{where the error variate } \epsilon_t \sim \boldsymbol{N} : 0, 1 \tag{5.5}$$

means that conditional upon $\boldsymbol{X}_{t-1} = x_{t-1}$ (and ϕ) the variate $\boldsymbol{X}_t \sim \boldsymbol{N} : \phi\, x_{t-1}, 1$. The model in Equation (5.5) is generally referred to as an autoregressive order one (AR(1)) model.

When observations exhibit first-order serial correlation, it means that the sample correlation between consecutively observed variates is non-negligible. That is, there is a pattern in the observed data that indicates a structure exists beyond what would be consistent with independent variates. However, this pattern may indeed be very useful if the prediction of future realized variates is a modeling objective.

To construct the joint distribution of the data, based on the assumption that the error variates $\boldsymbol{\epsilon}_t$, for $t = 1, 2, \ldots, n$ are *i.i.d* $\boldsymbol{N} : 0, 1$, we use the conditioning relationship implied by Equation (5.5) and the principle of composition discussed in Section 4.4. The joint density for the multivariate $\boldsymbol{X} = (\boldsymbol{X}_1, \boldsymbol{X}_2, \ldots, \boldsymbol{X}_n)$ can be described as the product of a marginal density and $n - 1$ successive conditional densities

$$f(x) = f(x_1) f(x_2|x_1) f(x_3|x_2, x_1) \cdots f(x_n|x_{n-1}, x_{n-2}, \ldots, x_1). \tag{5.6}$$

In this particular example and due to the AR(1) model structure, for $t \geq 2$, the densities in Equation (5.6) have the same form

$$\begin{aligned} f(x_t|x_{t-1}, x_{t-2}, \ldots, x_1) &= f(x_t|x_{t-1}) \\ &= f_{\epsilon_t}(x_t - \phi\, x_{t-1}) \\ &= \frac{1}{(2\,\pi)^{1/2}} \exp\left(\frac{-(x_t - \phi\, x_{t-1})^2}{2}\right). \end{aligned}$$

The first term in Equation (5.6) is problematic for $|\phi| \geq 1$, however, for $|\phi| < 1$ the variate $\boldsymbol{X}_1 \sim \boldsymbol{N} : 0, (1 - \phi^2)^{-1}$. The joint density of the multivariate is then

$$\begin{aligned} f(x) &= \frac{\left(1 - \phi^2\right)^{1/2}}{(2\,\pi)^{1/2}} \exp\left(\frac{-\left(1 - \phi^2\right)\left(x_1^2\right)}{2}\right) \prod_{t=2}^{n} \frac{1}{(2\,\pi)^{1/2}} \exp\left(\frac{-(x_t - \phi\, x_{t-1})^2}{2}\right) \\ &= \frac{\left(1 - \phi^2\right)^{1/2}}{(2\,\pi)^{n/2}} \exp\left(-\frac{1}{2}\left[(1 - \phi^2)(x_1^2) + \sum_{t=2}^{n}(x_t - \phi\, x_{t-1})^2\right]\right). \end{aligned}$$

Chapter 38 details many results from dependent structures generated by queues. Readers interested in more complex time series dependency structures are directed to Box, Jenkins, and Reinsel (2008).

Chapter 6

Parameter Inference

6.1 INTRODUCTION

In this chapter we review common methods of inference, namely *the Method of Percentiles (MP)*, *the Method of Moments (MM)*, *Maximum Likelihood (ML)* inference, and *Bayesian (B)* inference. Here we focus only on parameter inference and, where possible, point and interval estimation. Inferential frameworks apply to both continuous and discrete variates and so here we refer here to both a probability density function for the continuous variate and the probability function for a discrete variate as simply a *probability function.* As before, integrals are used to denote either integration or summation, depending on whether the relevant argument has continuous or discrete support.

The purpose of this chapter is to summarize the basic ideas behind each inferential method and to facilitate the use of the distributions provided in the latter portion of this book. Many more pages could have been devoted to each of the methods summarized here, but instead more detailed discussions already available are recommended for further reading. In addition, the book by Cox (2006) provides a lucid overview of the fundamental principles of statistical inference from a relatively pragmatic point of view.

6.2 METHOD OF PERCENTILES ESTIMATION

The idea behind the method of percentiles is simply to match the quantiles of the empirical distribution function, $F_E(x)$, to the theoretical quantile x_Y implied by a particular distribution function $F_{\boldsymbol{\theta}}(x)$. Here the subscript $\boldsymbol{\theta}$ emphasizes that the distribution function (previously denoted by $F_X(x)$ or simply $F(x)$) depends upon the value of the potentially vector valued parameter $\boldsymbol{\theta}$.

Consider the case where observations are *i.i.d.* but with ordered realized variates $x_1 \leq x_2 \leq \cdots \leq x_n$. Then x_i denotes the (i/n)th empirical quantile, or equivalently, the $100(i/n)$th percentile. In the scalar case, the method of percentiles finds a

Statistical Distributions, Fourth Edition, by Catherine Forbes, Merran Evans, Nicholas Hastings, and Brian Peacock

parameter value $\widehat{\boldsymbol{\theta}}_{MP}$ for a selected distribution $F_{\boldsymbol{\theta}}(x)$ such that when F is evaluated at the empirical quantile x_i, its value is equal to i/n. That is, $F_{\widehat{\boldsymbol{\theta}}_{MP}}(x_i) = 1/n$. This is equivalent to finding $\widehat{\boldsymbol{\theta}}_{MP}$ such that the corresponding inverse distribution function satisfies $G_{\widehat{\boldsymbol{\theta}}_{MP}}(i/n) = x_i$.

EXAMPLE 6.2.1. ***Exponential Distribution***

In this example θ is the scale parameter b. (See Chapter 17.) To find $\widehat{\boldsymbol{\theta}}_{MP} = \widehat{b}$ for the $\boldsymbol{E}: b$ we first note that

$$F_b(x) = 1 - \exp(-x/b)$$

$$G_b(\alpha) = -b\,log(1 - \alpha).$$

Using the inverse distribution function associated with the probability i/n, we find $\widehat{b}$ such that

$$G_{\widehat{b}}(i/n) = x_i,$$

resulting in

$$\widehat{b} = x_i \log\left(\frac{n}{n - i}\right).$$

Note that in this case the method may be applied to any of the empirical quantiles, except for the last associated with x_n.

The approach is relatively easy to implement, even if analytical expression for the inverse function G is not available. When the specified distribution involves more than one parameter, solving a set of equations based on more than one empirical quantile will produce the parameter estimates. Note the approach is easily adapted to the case of censoring; see Chapter 14.

Despite its intuitive appeal, the method of percentiles approach does not necessarily produce estimators with good statistical properties. It may, however, provide a suitable starting value for other iterative estimation methods.

6.3 METHOD OF MOMENTS ESTIMATION

Method of moments estimation seeks to find parameter values that set theoretical (population) moments equal to the corresponding sample moments. Suppose the distribution of interest depends upon a k-dimensional parameter vector, $\boldsymbol{\theta}$, and that the first k moments about the origin of the distribution, μ'_r, for $r = 1, 2, \ldots, k$, exist and are finite. To emphasize that theoretically these moments depend upon $\boldsymbol{\theta}$, here we denote the rth moment as $\mu'_r(\boldsymbol{\theta})$. Let $m'_r = \frac{1}{n}\sum_{i=1}^{n} x_i^r$ denote the rth sample moment about the origin. A method of moments estimator $\widehat{\boldsymbol{\theta}}_{MM}$ is simply a value of $\boldsymbol{\theta}$ simultaneously satisfying each of the k equations

$$\mu'_r(\widehat{\boldsymbol{\theta}}_{MM}) = m'_r, \quad \text{for } r = 1, 2, \ldots k. \tag{6.1}$$

Note that Equation (6.1) is not the only set of moment equations that produce method of moments estimator. For example, let m_r denote the rth (unbiased) sample moment about the about the sample mean $m_1' = \overline{x}$. An alternative method of moments estimator $\widetilde{\boldsymbol{\theta}}_{MM}$ is found by simultaneously solving

$$\mu_r(\widetilde{\boldsymbol{\theta}}_{MM}) = m_r, \quad \text{for } r = 1, 2, \ldots k. \tag{6.2}$$

Of course if $k = 1$ the two expressions in Equations (6.1) and (6.2) would yield the same result. However, even in that case one could choose to find a $\widehat{\theta}_{MM}$ using a higher order sample moment, or, for example,

$$\int_{-\infty}^{\infty} g_{\theta}(x)\, f_{\theta}(x)\, dx = \frac{1}{n}\sum_{i=1}^{n} g_{\theta}(x_i).$$

Extensions to alternative moment estimators for the $k > 1$ dimensional vector $\boldsymbol{\theta}$ case are also possible.

EXAMPLE 6.3.1. ***Gamma Distribution***

The gamma distribution has two parameter, so $\boldsymbol{\theta}' = (b, c)$, with mean $\mu_1'(\boldsymbol{\theta}) = bc$ and variance $\mu_2(\boldsymbol{\theta}) = b^2c$. Also, note that the corresponding first two (unbiased) sample moments are $m_1 = \overline{x}$ and $m_2 = s_u^2$, where s_u^2 is defined in Table 2.3. Solving for the values of $\hat{b}$ and $\hat{c}$ such that

$$\hat{b}\,\hat{c} = \overline{x}$$
$$\hat{b}^2\,\hat{c} = s_u^2$$

results in

$$\hat{b} = s_u^2/\overline{x}$$
$$\widehat{c} = \overline{x}^2/s_u^2.$$

Method of moments estimators can have desirable statistical properties, such as *consistency*. Loosely speaking, an estimator is consistent if the chance of it being very close to the true value increases to one as the sample size increases. That is, an estimator $\widehat{\boldsymbol{\theta}}_n$ is consistent if it converges in probability to the true parameter $\boldsymbol{\theta}$. More formally

$$\lim_{n\to\infty} \Pr(|\widehat{\boldsymbol{\theta}}_n - \boldsymbol{\theta}| < \epsilon) = 1 \tag{6.3}$$

for any small $\epsilon > 0$.

Extensions of this approach, known in the econometrics literature as *Generalized Method of Moments (GMM)* and *Estimating Equations* in the statistics literature, have become popular in recent times, particularly for when modeling dependent data or where weaker assumptions regarding the underlying distribution are imposed. For more details and GMM examples, we refer readers to Johnston and DiNardo (1997), with more technical details available in Hall (2005).

6.4 MAXIMUM LIKELIHOOD INFERENCE

When the joint density for a set of variates is viewed as a function of the parameters alone, rather than as a function of the variates themselves, that function is called a *likelihood function*. Hence the likelihood function, $L(\boldsymbol{\theta})$, is defined as

$$L(\boldsymbol{\theta}) = f_{\boldsymbol{\theta}}(\boldsymbol{x})$$

and in particular when variates are *i.i.d.*

$$L(\boldsymbol{\theta}) = \prod_{i=1}^{n} f_{\boldsymbol{\theta}}(x_i).$$

Here $\log f_{\boldsymbol{\theta}}(x)$ is a scalar function of a k-dimensional variable $\boldsymbol{\theta}$ and as in Chapter 5 $x = (x_1, x_2, \ldots, x_n)$. A value of the parameter $\boldsymbol{\theta}$ that maximizes $L(\boldsymbol{\theta})$ is called a maximum likelihood estimator (MLE), and is denoted by $\widehat{\boldsymbol{\theta}}_{ML}$. It is often easier to maximize the *log-likelihood function*, $\log L(\boldsymbol{\theta})$, and since the (natural) logarithmic function is monotonically increasing in $\boldsymbol{\theta}$, the same value of $\widehat{\boldsymbol{\theta}}_{ML}$ maximizes both $L(\boldsymbol{\theta})$ and $\log L(\boldsymbol{\theta})$.

Properties of MLEs

Under quite general conditions, MLEs have a number of favorable properties. We consider here those associated with *i.i.d.* variates, although generalizations of these properties hold for a likelihood function based on dependent variates in many settings. These properties essentially all have to do with the behavior of MLEs when the sample size is very large. We place an additional subscript 'n' on $\widehat{\boldsymbol{\theta}}_{ML}$ to emphasize the fact that any particular MLE will be a function of the sample size, n.

Consistency. Under mild conditions, MLEs converge to the true parameter value as the sample size increases. See Equation (6.3).

Asymptotic Normality. As the sample size increases, the distribution of the MLE approaches that of a (potentially) Multivariate Normal variate. In particular,

$$\sqrt{n}\left(\widehat{\boldsymbol{\theta}}_{ML,n} - \boldsymbol{\theta}\right) \overset{Dist}{\to} MN : \mathbf{0},\ \boldsymbol{i}(\boldsymbol{\theta})^{-1}.$$

Here $\mathbf{0}$ denotes the k-dimensional zero vector. The $(k \times k)$ dimensional matrix $\boldsymbol{i}(\boldsymbol{\theta})$ is called the *(Fisher) information*, and is equal to minus the expected second partial derivative matrix of the log-probability function of a single variate, evaluated at the "true" parameter, $\boldsymbol{\theta}$

$$\boldsymbol{i}(\boldsymbol{\theta}) = -E\left[\frac{\partial^2 \log f_{\boldsymbol{\theta}}(x)}{\partial\boldsymbol{\theta}\partial\boldsymbol{\theta}'}\right] = -\int_{-\infty}^{\infty}\left[\frac{\partial^2 \log f(x|\boldsymbol{\theta})}{\partial\boldsymbol{\theta}\partial\boldsymbol{\theta}'}\right] f_{\boldsymbol{\theta}}(x)\,dx. \qquad (6.4)$$

In the multidimensional setting, $\partial \log f_{\boldsymbol{\theta}}(\boldsymbol{x})/\partial\boldsymbol{\theta}$ denotes a column vector of partial derivatives with $\partial \log f_{\boldsymbol{\theta}}(\boldsymbol{x})/\partial\theta_r$ in the rth row, θ_r denotes the rth component of $\boldsymbol{\theta}$, and $\partial^2 \log f(\boldsymbol{x}|\boldsymbol{\theta})/\partial\boldsymbol{\theta}\partial\boldsymbol{\theta}'$ denotes the matrix of second order partial derivatives with row (r) and column (c) element $\partial^2 \log f_{\boldsymbol{\theta}}(\boldsymbol{x})/\partial\theta_r\partial\theta_c$.

Asymptotic Efficiency. MLEs are asymptotically efficient, meaning that even if an alternative estimator, $\widetilde{\boldsymbol{\theta}}_n$, is available, with

$$\sqrt{n}\left(\widetilde{\boldsymbol{\theta}}_n - \boldsymbol{\theta}\right) \overset{Dist}{\rightarrow} MN : \mathbf{0},\ \boldsymbol{\delta},$$

then always $\boldsymbol{\delta} \geq \boldsymbol{i}(\boldsymbol{\theta})^{-1}$. Here $\overset{Dist}{\rightarrow}$ denotes *convergence in distribution*, meaning that as the sample size increases then the distribution of the (appropriately scaled and shifted) parameter estimator $\widetilde{\boldsymbol{\theta}}_n$ behaves as a $MN : \mathbf{0},\ \boldsymbol{\delta}$ variate.

Invariance. Under mild conditions, the MLE of a differentiable function $\boldsymbol{g}(\boldsymbol{\theta})$, is equal to that function evaluated at the MLE of $\boldsymbol{\theta}$. That is, if $\widehat{\boldsymbol{\theta}}_{ML}$ is the MLE of $\boldsymbol{\theta}$, then $\boldsymbol{g}\left(\widehat{\boldsymbol{\theta}}_{ML}\right)$ is the MLE of $\boldsymbol{g}(\boldsymbol{\theta})$, and further

$$\sqrt{n}\left(\boldsymbol{g}\left(\widehat{\boldsymbol{\theta}}_{ML,n}\right) - \boldsymbol{g}(\boldsymbol{\theta})\right) \overset{Dist}{\rightarrow} MN : \mathbf{0},\ \left[\frac{\partial \boldsymbol{g}(\boldsymbol{\theta})}{\partial \boldsymbol{\theta}}\right] \boldsymbol{i}(\boldsymbol{\theta})^{-1} \left[\frac{\partial \boldsymbol{g}(\boldsymbol{\theta})}{\partial \boldsymbol{\theta}}\right]'.$$

Since $\boldsymbol{g}(\boldsymbol{\theta})$ may be a $g \leq k$ dimensional vector function of a k-dimensional vector argument, then $\left[\partial \boldsymbol{g}(\boldsymbol{\theta})/\partial\boldsymbol{\theta}\right]$ represents the matrix of partial derivatives with $[r, c]$ element $\left[\partial g_r(\boldsymbol{\theta})/\partial\theta_c\right]$, where $g_r(\boldsymbol{\theta})$ denotes the rth scalar function element of $\boldsymbol{g}(\boldsymbol{\theta})$. $\left[\partial \boldsymbol{g}(\boldsymbol{\theta})/\partial\boldsymbol{\theta}\right]'$ denotes the transpose of $\left[\partial \boldsymbol{g}(\boldsymbol{\theta})/\partial\boldsymbol{\theta}\right]$.

Approximate Sampling Distribution for Fixed n

To make use of the asymptotic normality result and its associated properties, the information matrix $\boldsymbol{i}(\boldsymbol{\theta})$ will commonly need to be estimated and this will usually be done in two ways. The first involves using the sample average of second partial derivative matrix of the log-probability function of each variate, evaluated at the MLE

$$\widehat{\boldsymbol{i}}_1 = -\frac{1}{n}\sum_{i=1}^{n} \left.\frac{\partial^2 \log f(x_i|\boldsymbol{\theta})}{\partial\boldsymbol{\theta}\partial\boldsymbol{\theta}'}\right|_{\boldsymbol{\theta}=\widehat{\boldsymbol{\theta}}_{ML}}.$$

The second involves the sample average of the outer product of the first derivative of the log-probability function of each variate, evaluated at the MLE

$$\widehat{\boldsymbol{i}}_2 = \frac{1}{n}\sum_{i=1}^{n} \left.\left(\frac{\partial \log f(x_i|\boldsymbol{\theta})}{\partial\boldsymbol{\theta}}\right)\left(\frac{\partial \log f(x_i|\boldsymbol{\theta})}{\partial\boldsymbol{\theta}}\right)'\right|_{\boldsymbol{\theta}=\widehat{\boldsymbol{\theta}}_{ML}}.$$

In either case, the resulting approximation for the sampling distribution for the MLE is $\widehat{\boldsymbol{\theta}}_{ML} \overset{approx}{\sim} MN : \boldsymbol{\theta},\ \left(n^{-1}\widehat{\boldsymbol{i}}^{-1}\right)$. A corresponding approximate sampling distribution for $g\left(\widehat{\boldsymbol{\theta}}_{ML}\right)$ is

$$\boldsymbol{g}\left(\widehat{\boldsymbol{\theta}}_{ML}\right) \overset{approx}{\sim} MN : \boldsymbol{g}(\boldsymbol{\theta}),\, n^{-1}\left[\frac{\partial \boldsymbol{g}\left(\widehat{\boldsymbol{\theta}}_{ML}\right)}{\partial \boldsymbol{\theta}}\right]\widehat{\boldsymbol{i}}^{-1}\left[\frac{\partial \boldsymbol{g}\left(\widehat{\boldsymbol{\theta}}_{ML}\right)}{\partial \boldsymbol{\theta}}\right]',$$

where $\partial \boldsymbol{g}\left(\widehat{\boldsymbol{\theta}}_{ML}\right)/\partial \boldsymbol{\theta}$ denotes $\partial \boldsymbol{g}(\boldsymbol{\theta})/\partial \boldsymbol{\theta}$ evaluated at $\widehat{\boldsymbol{\theta}}_{ML}$.

EXAMPLE 6.4.1. ***Exponential Sample***

Assume $\boldsymbol{X}_1, \boldsymbol{X}_2, \ldots, \boldsymbol{X}_n$ are *i.i.d.*$\sim$ $\boldsymbol{E} : b$. Then the log-likelihood function is

$$\log L(b) = \log\left[\prod_{i=1}^{n} \frac{1}{b}\exp(-x_i/b)\right] = -n\log b - n\overline{x}/b.$$

Differentiating the log-likelihood (with respect to b) yields

$$\frac{d\log L(b)}{db} = -nb^{-1} + n\overline{x}b^{-2}.$$

The MLE is found by setting this first derivative to zero, that is,

$$-n\hat{b}_{ML}^{-1} + n\overline{x}\hat{b}_{ML}^{-2} = 0$$

and solving for $\hat{b}_{ML} = \overline{x}$. Since $\overline{x} > 0$, the second derivative of $\log L(b)$ evaluated at $\hat{b}_{ML} = -n(\overline{x})^{-2}$ is negative, confirming that $\hat{b}_{ML}$ maximizes the likelihood function.

Since b is a scalar parameter, the information is a scalar function of b and is found by twice differentiating $\log f_b(x) = -\log b - x/b$. In this case,

$$i(b) = -E\left[\frac{d^2\log f(x|b)}{db^2}\right] = b^{-2}$$

yielding the result (for fixed n) that

$$\overline{x} \overset{\text{approx}}{\sim} N : b, b^2/n.$$

An approximate $100(1-\alpha)\%$ confidence interval for b is given by

$$\overline{x} \pm z_{\alpha/2}\overline{x}/\sqrt{n}$$

where $z_{d/2} = G_{N:0,1}(1-\alpha/2)$ denotes the $(1-\alpha/2)100\%$ quantile of the $\boldsymbol{N} : 0, 1$ distribution.

The above example is simple enough to permit analytical calculation of the MLE and its associated information. However, in many applications the MLE and the information matrix must be computed numerically on a computer. For more details on the application of maximum likelihood in practice, or for more background on the theoretical properties involved, see Casella and Berger (2002).

6.5 BAYESIAN INFERENCE

For Bayesian analysis of statistical problems, probability statements are considered a measure of belief. Bayesian statistics involves the formal updating of prior (pre-data) beliefs about parameter values in light of observed information (data) through a formal revision of probabilities undertaken using Bayes' theorem. (See Section 4.5.) In addition to one's subjective view, it is common to summarize prior belief arising from a range of sources, including thought experiments, knowledge of previous analysis of similar data, and consensus opinion.

In this context parameters are treated as though they are random variates with prior belief regarding parameter values summarized in a prior probability function $f(\boldsymbol{\theta})$. This prior probability function is then revised in light of the likelihood function via Bayes theorem to produce an updated probability function for $\boldsymbol{\theta}$, called the *posterior probability function*

$$f(\boldsymbol{\theta}|\boldsymbol{x}) = \frac{f(\boldsymbol{\theta})\, L(\boldsymbol{\theta})}{m(\boldsymbol{x})}. \tag{6.5}$$

The function $m(\boldsymbol{x})$ in the denominator of Equation (6.5) denotes the so-called *marginal likelihood,* with

$$m(\boldsymbol{x}) = \int f(\boldsymbol{\theta})\, L(\boldsymbol{\theta})\, d\boldsymbol{\theta}. \tag{6.6}$$

The value of $m(\boldsymbol{x})$ is found by evaluating the marginal probability function of $\boldsymbol{X}$ in Equation (6.6) at the observed $\boldsymbol{X} = \boldsymbol{x}$. Since it is not a function of $\boldsymbol{\theta}$ and x is known, $m(\boldsymbol{x})$ is constant. As before $L(\boldsymbol{\theta})$ represents the joint *probability function* of the observed data viewed as a function of the unknown parameter, $\boldsymbol{\theta}$. To emphasize that the distribution of the multivariate $\boldsymbol{X}$ is known conditionally upon a fixed value of the parameter $\boldsymbol{\theta}$, the probability function $f_{\boldsymbol{\theta}}(\boldsymbol{x})$ is represented using the conditioning notation $f(\boldsymbol{x}|\boldsymbol{\theta})$.

The posterior probability function provides the mechanism for computing probabilities associated with the posterior distribution. Hence the posterior distribution itself is the complete summary for the unknown parameter. Once obtained, the expected value (when it exists)

$$\widehat{\boldsymbol{\theta}}_B = \int_{-\infty}^{\infty} \boldsymbol{\theta}\, f(\boldsymbol{\theta}|\boldsymbol{x})\, d\boldsymbol{\theta}$$

and the mode of the posterior

$$\widetilde{\boldsymbol{\theta}}_B = \max_{\boldsymbol{\theta}} f(\boldsymbol{\theta}|\boldsymbol{x})$$

are frequently cited as a representative value from this distribution and are sometimes regarded as point estimators for $\boldsymbol{\theta}$.

Marginal Posteriors

When $\boldsymbol{\theta}$ is multivariate, the distribution of the *rth* element of the parameter vector, θ_r, may be summarized using its marginal posterior distribution, found in the usual way by integrating over the remaining parameter components

$$f(\theta_r|\boldsymbol{x}) = \int f(\boldsymbol{\theta}|\boldsymbol{x})\, d\boldsymbol{\theta}_{(r)}$$

where $\boldsymbol{\theta}_{(r)}$ denotes all of the elements of $\boldsymbol{\theta}$ *except* the rth element, θ_r.

EXAMPLE 6.5.1. ***Exponential Variates with a Gamma Prior***

Consider *i.i.d.* univariates $\boldsymbol{X}_1, \boldsymbol{X}_2, \ldots, \boldsymbol{X}_n \sim \boldsymbol{E} : \lambda$ with prior distribution $\lambda \sim \boldsymbol{\gamma} : b, c$. That is, the prior distribution of the parameter λ is that of a Gamma variate with mean bc, having a prior mode of $b(c-1)$, and with prior probability function for $\lambda > 0$ given by

$$f(\lambda) = \frac{\lambda^{c-1}}{b^c\,\Gamma(c)} \exp\left\{-\frac{\lambda}{b}\right\}.$$

The likelihood function for λ associated with the multivariate $\boldsymbol{X}$ is

$$L(\lambda) = \prod_{i=1}^{n} \lambda \exp\{-\lambda x_i\} = \lambda^n \exp\{-\lambda n\bar{x}\}$$

so that the posterior probability function has the form

$$\begin{aligned} f(\lambda|\boldsymbol{x}) &= \lambda^n \exp\{-\lambda n\bar{x}\} \frac{\lambda^{c-1}}{b^c\Gamma(c)} \exp\left\{-\frac{\lambda}{b}\right\} / m(\boldsymbol{x}) \\ &= \frac{\lambda^{n+c-1}}{b^c\Gamma(c)} \exp\left\{-\lambda\left[n\bar{x} + \frac{1}{b}\right]\right\} / m(\boldsymbol{x}). \end{aligned}$$

Since

$$\begin{aligned} m(\boldsymbol{x}) &= \int_0^\infty \frac{\lambda^{n+c-1}}{b^c\Gamma(c)} \exp\left\{-\lambda\left[n\bar{x} + \frac{1}{b}\right]\right\} d\lambda \\ &= \frac{\Gamma(n+c)}{b^c\Gamma(c)\left[n\bar{x} + \frac{1}{b}\right]^{n+c}} \\ &\quad \times \int_0^\infty \frac{\left[n\bar{x} + \frac{1}{b}\right]^{n+c}}{\Gamma(n+c)} \lambda^{n+c-1} \exp\left\{-\lambda\left[n\bar{x} + \frac{1}{b}\right]\right\} d\lambda \\ &= \frac{b^n\,\Gamma(n+c)}{\Gamma(c)\,[bn\bar{x} + 1]^{n+c}} \end{aligned}$$

the posterior probability function is

$$f(\lambda|\boldsymbol{x}) = \left[\frac{bn\bar{x} + 1}{b}\right]^{n+c} \frac{1}{\Gamma(n+c)} \lambda^{n+c-1} \exp\left\{-\lambda\left[n\bar{x} + \frac{1}{b}\right]\right\}, \qquad (6.7)$$

implying

$$\lambda | \boldsymbol{x} = \boldsymbol{\gamma} : b[b n \overline{x} + 1]^{-1}, (n + c).$$

Therefore the posterior distribution has a single mode at

$$\hat{\lambda}_M = b(n + c - 1)/(b n \overline{x} + 1)$$

and its mean is

$$\hat{\lambda}_B = b(n + c)/(b n \overline{x} + 1).$$

The shortest available $100\,(1 - \alpha)\%$ credibility interval (L_λ, U_λ) would be found such that

$$\begin{aligned} 0.95 &= Pr(L_\lambda < \lambda \leq U_\lambda | \boldsymbol{X}) \\ &= \int_{L_\lambda}^{U_\lambda} f(\lambda | \boldsymbol{x})\, d\lambda \end{aligned}$$

where $f(\lambda | \boldsymbol{x})$ is shown in Equation (6.7).

Often the most difficult part of a Bayesian analysis is computing the marginal likelihood $m(\boldsymbol{x})$. However, in recent years the vast improvements in computing capacity and the development of simulation methods have meant that once intractable problems may now be numerically solved. In particular, empirical analyses using Bayesian methods frequently use so-called *Markov chain Monte Carlo (MCMC)* methods to draw a sample of parameter values $\boldsymbol{\theta}$ from its posterior distribution and to produce point and interval summaries of that posterior distribution. For a detailed exposition including both philosophical and methodological developments, see Robert (2007).

In small dimensional and regular problems, as the sample size increases there is often little difference between the MLE and Bayesian point estimators. However, in problems with a large set of unknown parameters, the two methods can differ significantly even for relatively large samples.

Chapter 7

Bernoulli Distribution

A Bernoulli trial is a probabilistic experiment that can have one of two outcomes, success ($x = 1$) or failure ($x = 0$), and in which the probability of success is p. We refer to p as the Bernoulli probability parameter.

An example of a Bernoulli trial is the inspection of a random item from a production line with the possible result that the item could be acceptable or faulty. The Bernoulli trial is a basic building block for other discrete distributions such as the binomial, Pascal, geometric, and negative binomial.

Variate B: 1, p.

(The general binomial variate is B: n, p, involving n trials.)

Range $x \in \{0, 1\}$.

Parameter p, the Bernoulli probability parameter, $0 < p < 1$.

Distribution function	$F(0) = 1 - p$; $F(1) = 1$
Probability function	$f(0) = 1 - p$; $f(1) = p$
Characteristic function	$1 + p[\exp(it) - 1]$
rth Moment about the origin	p
Mean	p
Variance	$p(1 - p)$

7.1 RANDOM NUMBER GENERATION

$\boldsymbol{R}$ is a unit rectangular variate and $\boldsymbol{B}$: 1, p is a Bernoulli variate. $\mathbf{R} \leq p$ implies $\boldsymbol{B}$: 1, p takes value 1; $\mathbf{R} > p$ implies $\boldsymbol{B}$: 1, p takes value 0.

7.2 CURTAILED BERNOULLI TRIAL SEQUENCES

The binomial, geometric, Pascal, and negative binomial variates are based on sequences of independent Bernoulli trials, which are curtailed in various ways, for example, after n trials or x successes. We shall use the following terminology:

Statistical Distributions, Fourth Edition, by Catherine Forbes, Merran Evans, Nicholas Hastings, and Brian Peacock

p	Bernoulli probability parameter (probability of success at a single trial)
n	Number of trials
x	Number of successes
y	Number of failures
$\boldsymbol{B}$: n, p	Binomial variate, number of successes in n trials
$\boldsymbol{G}$: p	Geometric variate, number of failures before the first success
$\boldsymbol{NB}$: x, p	Negative binomial variate, number of failures before the xth success

A Pascal variate is the integer version of the negative binomial variate. Alternative forms of the geometric and Pascal variates include the number of trials up to and including the xth success. These variates are interrelated in various ways, specified under the relevant chapter headings.

7.3 URN SAMPLING SCHEME

The selection of items from an urn, with a finite population N of which Np are of the desired type or attribute and $N(1-p)$ are not, is the basis of the Polyà family of distributions.

A Bernoulli variate corresponds to selecting one item ($n = 1$) with probability p of success in choosing the desired type. For a sample consisting of n independent selections of items, with replacement, the binomial variate $\boldsymbol{B}$: n, p is the number x of desired items chosen or successes, and the negative binomial variate $\boldsymbol{NB}$: x, p is the number of failures before the xth success. As the number of trials or selections n tends to infinity, p tends to zero, and np tends to a constant λ, the binomial variate tends to the Poisson variate $\boldsymbol{P}$: λ with parameter $\lambda = np$.

If sample selection is without replacement, successive selections are not independent, and the number of successes x in n trials is a hypergeometric variate $\boldsymbol{H}$: N, x, n. If two items of the type corresponding to that selected are replaced each time, thus introducing "contagion," the number of successes x in n trials is then a negative hypergeometric variate, with parameters N, x, and n.

7.4 NOTE

The following properties can be used as a guide in choosing between the binomial, negative binomial, and Poisson distribution models:

Binomial	Variance $<$ mean
Negative binomial	Variance $>$ mean
Poisson	Variance $=$ mean

Chapter 8

Beta Distribution

Applications include modeling random variables that have a finite range, a to b. An example is the distribution of activity times in project networks. The beta distribution is frequently used as a prior distribution for binomial proportions in Bayesian analysis.

Variate $\boldsymbol{\beta} : \nu,\ \omega$.

Range $0 \le x \le 1$.

Shape parameters $\nu > 0, \omega > 0$.

This beta distribution (of the first kind) is U shaped if $\nu < 1,\ \omega < 1$ and J shaped if $(\nu - 1)(\omega - 1) < 0$, and is otherwise unimodal.

Distribution function	Often called the incomplete beta function. (See Pearson, 1968.)
Probability density function	$x^{\nu-1}(1-x)^{\omega-1}/\mathrm{B}(\nu,\ \omega)$, where $\mathrm{B}(\nu,\ \omega)$ is the beta function with arguments $\nu,\ \omega$, given by $\mathrm{B}(\nu,\ \omega) = \int_0^1 u^{\nu-1}(1-u)^{\omega-1}du$
rth Moment about the origin	$\prod_{i=0}^{r-1} \dfrac{(\nu+i)}{(\nu+\omega+i)} = \dfrac{\mathrm{B}(\nu+r,\ \omega)}{\mathrm{B}(\nu,\ \omega)}$
Mean	$\nu/(\nu+\omega)$
Variance	$\nu\omega/\left[(\nu+\omega)^2(\nu+\omega+1)\right]$
Mode	$(\nu-1)/(\nu+\omega-2),\ \nu > 1, \omega > 1$
Coefficient of skewness	$\dfrac{2(\omega-\nu)(\nu+\omega+1)^{1/2}}{(\nu+\omega+2)(\nu\omega)^{1/2}}$

Statistical Distributions, Fourth Edition, by Catherine Forbes, Merran Evans, Nicholas Hastings, and Brian Peacock

Coefficient of kurtosis	$\frac{3(\nu+\omega)(\nu+\omega+1)(\nu+1)(2\omega-\nu)}{\nu\omega(\nu+\omega+2)(\nu+\omega+3)} + \frac{3(\nu+\omega+1)(\nu-\omega)}{\omega(\nu+\omega+2)}$
Coefficient of variation	$\left(\frac{\omega}{\nu(\nu+\omega+1)}\right)^{1/2}$
Probability density function if ν and ω are integers	$\frac{(\nu+\omega-1)!x^{\nu-1}(1-x)^{\omega-1}}{(\nu-1)!(\omega-1)!}$
Probability density function if range is $a \le x \le b$. Here a is a location parameter and $b-a$ a scale parameter.	$\frac{(x-a)^{\nu-1}(b-x)^{\omega-1}}{B(\nu,\omega)(b-a)^{\nu+\omega-1}}$.

8.1 NOTES ON BETA AND GAMMA FUNCTIONS

The beta function with arguments ν, ω is denoted $B(\nu, \omega)$; $\nu, \omega > 0$. The gamma function with argument c is denoted $\Gamma(c)$; $c > 0$. The di-gamma function with argument c is denoted $\psi(c)$; $c > 0$.

Definitions

Beta function:

$$B(\nu, \omega) = \int_0^1 u^{\nu-1}(1-u)^{\omega-1}\, du$$

Gamma function:

$$\Gamma(c) = \int_0^\infty \exp(-u)u^{c-1}\, du$$

Di-gamma function:

$$\psi(c) = \frac{d}{dc}\left[\log \Gamma(c)\right] = \frac{d\Gamma(c)/dc}{\Gamma(c)}$$

Interrelationships

$$B(\nu, \omega) = \frac{\Gamma(\nu)\Gamma(\omega)}{\Gamma(\nu+\omega)} = B(\omega, \nu)$$

$$\Gamma(c) = (c-1)\Gamma(c-1)$$

$$B(\nu+1, \omega) = \frac{\nu}{\nu+\omega}B(\nu, \omega)$$

Special Values

If ν, ω, and c are integers,

$$B(\nu, \omega) = (\nu - 1)!(\omega - 1)!/(\nu + \omega - 1)!$$

$$\Gamma(c) = (c - 1)!$$

$$B(1, 1) = 1, \quad B\left(\tfrac{1}{2}, \tfrac{1}{2}\right) = \pi$$

$$\Gamma(1) = 1, \quad \Gamma\left(\tfrac{1}{2}\right) = \pi^{1/2}, \quad \Gamma(2) = 1$$

Alternative Expressions

$$\begin{aligned} B(\nu, \omega) &= 2\int_0^{\pi/2} \sin^{2\nu-1}\theta \cos^{2\omega-1}\theta \, d\theta \\ &= \int_0^{\infty} \frac{y^{\omega-1}\, dy}{(1+y)^{\nu-\omega}} \end{aligned}$$

8.2 VARIATE RELATIONSHIPS

For the range $a \le x \le b$, the beta variate with parameters ν and ω is related to the beta variate with the same shape parameters but with the range $0 \le x \le 1$, $(\boldsymbol{\beta} : \nu, \omega)$ by

$$b(\boldsymbol{\beta} : \nu, \omega) + a\left[1 - (\boldsymbol{\beta} : \nu, \omega)\right].$$

1. The beta variates $\boldsymbol{\beta}$: ν, ω, and $\boldsymbol{\beta}$: ω, ν exhibit symmetry; see Figures 8.1 and 8.2. In terms of probability statements and the distribution functions, we have

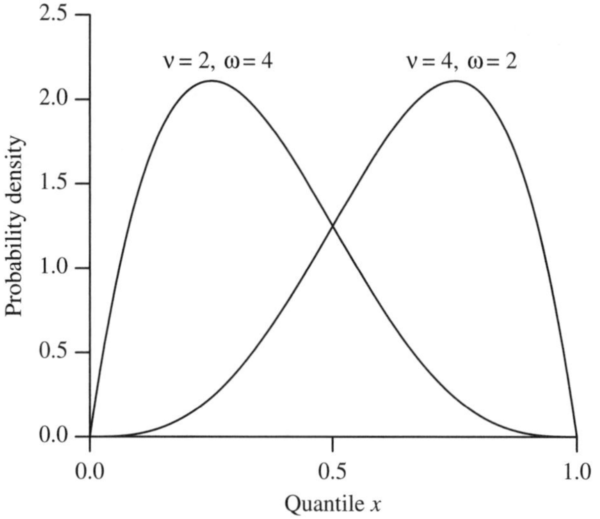

Figure 8.1. Probability density function for the beta variate $\boldsymbol{\beta} : \nu, \omega$.

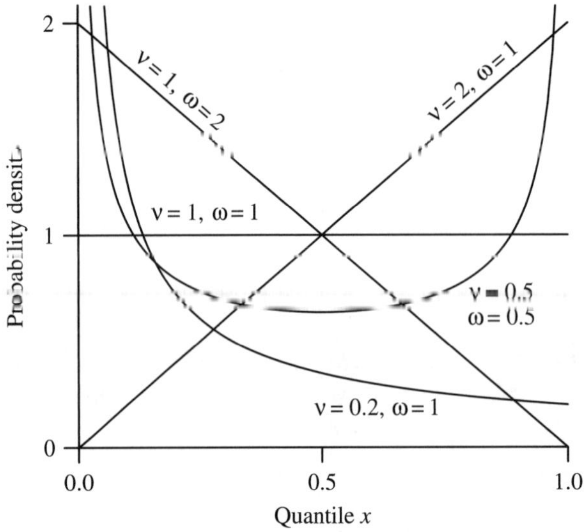

Figure 8.2. Probability density function for the beta variate $\boldsymbol{\beta}: \nu,\ \omega$ for additional values of the parameters.

$$\begin{aligned}\Pr\left[(\boldsymbol{\beta}: \nu,\ \omega) \le x\right] &= 1 - \Pr\left[(\boldsymbol{\beta}: \omega,\ \nu) \le (1-x)\right] \\ &= \Pr\left[(\boldsymbol{\beta}: \omega,\ \nu) > (1-x)\right] \\ &= F_{\boldsymbol{\beta}}(x : \nu,\ \omega) = 1 - F_{\boldsymbol{\beta}}\big((1-x) : \omega,\ \nu\big).\end{aligned}$$

2. The beta variate $\boldsymbol{\beta} : \frac{1}{2}, \frac{1}{2}$ is an arc sin variate (Figs. 8.2 and 8.3).

3. The beta variate $\boldsymbol{\beta}$: 1, 1 is a rectangular variate (Figs. 8.2 and 8.3).

4. The beta variate $\boldsymbol{\beta} : \nu$, 1 is a power function variate.

5. The beta variate with shape parameters $i, n-i+1$, denoted $\boldsymbol{\beta} : i, n-i+1$, and the binomial variate with Bernoulli trial parameter n and Bernoulli probability parameter p, denoted $\boldsymbol{B}: n,\ p$, are related by the following equivalent statements:

$$\begin{aligned}\Pr\left[(\boldsymbol{\beta} : i, n-i+1) \le p\right] &= \Pr\left[(\boldsymbol{B} : n,\ p) \ge i\right] \\ F_{\boldsymbol{\beta}}(p : i, n-i+1) &= 1 - F_{\boldsymbol{B}}(i-1 : n,\ p).\end{aligned}$$

Here n and i are positive integers, $0 \le p \le 1$.

Equivalently, putting $\nu = i,\ \omega = n-i+1$, and $x = p$:

$$\begin{aligned}F_{\boldsymbol{\beta}}(x : \nu,\ \omega) &= 1 - F_{\boldsymbol{B}}(\nu - 1 : \nu + \omega - 1, x) \\ &= F_{\boldsymbol{B}}(\omega - 1 : \nu + \omega - 1, 1 - x).\end{aligned}$$

6. The beta variate with shape parameters $\omega/2, \nu/2$, denoted $\boldsymbol{\beta} : \omega/2, \nu/2$, and the $\boldsymbol{F}$ variate with degrees of freedom ν, ω, denoted $\boldsymbol{F}: \nu,\ \omega$, are related by

$$\Pr\left[(\boldsymbol{\beta} : \omega/2, \nu/2) \le \omega/(\omega + \nu x)\right] = \Pr\left[(\boldsymbol{F} : \nu,\ \omega) > x\right].$$

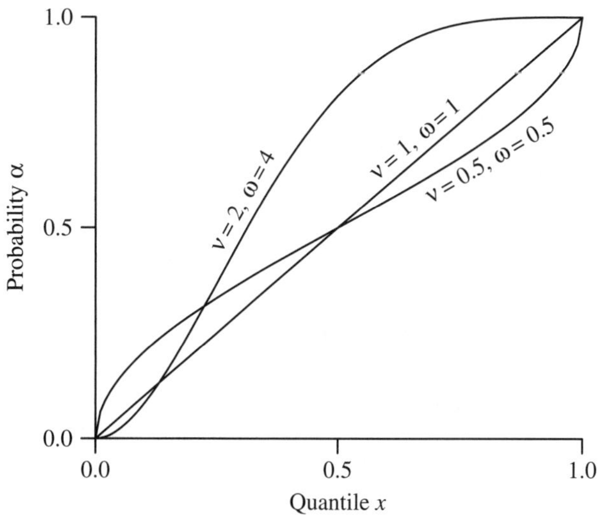

Figure 8.3. Distribution function for the beta variate $\boldsymbol{\beta} : \nu,\ \omega$.

Hence the inverse distribution function $G_{\boldsymbol{\beta}}(\alpha : \omega/2, \nu/2)$ of the beta variate $\boldsymbol{\beta} : \omega/2, \nu/2$ and the inverse survival function $Z_{\boldsymbol{F}}(\alpha : \nu, \omega)$ of the $\boldsymbol{F}$ variate $\boldsymbol{F}$: ν, ω are related by

$$\begin{aligned}(\omega/\nu)\{[1/G_{\boldsymbol{\beta}}(\alpha : \omega/2, \nu/2)] - 1\} &= Z_{\boldsymbol{F}}(\alpha : \nu,\ \omega) \\ &= G_{\boldsymbol{F}}(1 - \alpha : \nu, \omega)\end{aligned}$$

where α denotes probability.

7. The independent gamma variates with unit scale parameter and shape parameter ν, denoted $\boldsymbol{\gamma} : 1, \nu$, and with shape parameter ω, denoted $\boldsymbol{\gamma} : 1, \omega$, respectively, are related to the beta variate $\boldsymbol{\beta} : \nu,\ \omega$ by

$$\boldsymbol{\beta} : \nu,\ \omega \sim (\boldsymbol{\gamma} : 1, \nu)/[(\boldsymbol{\gamma} : 1, \nu) + (\boldsymbol{\gamma} : 1, \omega)].$$

8. As ν and ω tend to infinity, such that the ratio ν/ω remains constant, the $\boldsymbol{\beta} : \nu,\ \omega$ variate tends to the standard normal variate $\boldsymbol{N}$: 0, 1.

9. The variate $\boldsymbol{\beta} : \nu,\ \omega$ corresponds to the one-dimensional Dirichlet variate with $\nu = c_1, \omega = c_0$. The Dirichlet distribution is the multivariate generalization of the beta distribution.

8.3 PARAMETER ESTIMATION

Parameter	Estimator	Method
ν	$\bar{x}\{[\bar{x}(1 - \bar{x})/s^2] - 1\}$	Matching moments
ω	$(1 - \bar{x})\{[\bar{x}(1 - \bar{x})/s^2] - 1\}$	Matching moments

The maximum-likelihood estimators $\hat{\nu}$ and $\hat{\omega}$ are the solutions of the simultaneous equations

$$\psi(\hat{\nu}) - \psi(\hat{\nu} + \hat{\omega}) = n^{-1} \sum_{i=1}^{n} \log x_i$$

$$\psi(\hat{\omega}) - \psi(\hat{\nu} + \hat{\omega}) = n^{-1} \sum_{i=1}^{n} \log(1 - x_i).$$

8.4 RANDOM NUMBER GENERATION

If ν and ω are integers, then random numbers of the beta variate $\boldsymbol{\beta} : \nu, \omega$ can be computed from random numbers of the unit rectangular variate $\boldsymbol{R}$ using the relationship with the gamma variates $\boldsymbol{\gamma} : 1, \nu$ and $\boldsymbol{\gamma} : 1, \omega$ as follows:

$$\boldsymbol{\gamma} : 1, \nu \sim -\log \prod_{i=1}^{\nu} \boldsymbol{R}_i$$

$$\boldsymbol{\gamma} : 1, \omega \sim -\log \prod_{j=1}^{\omega} \boldsymbol{R}_i$$

$$\boldsymbol{\beta} : \nu, \omega \sim \frac{\boldsymbol{\gamma} : 1, \nu}{(\boldsymbol{\gamma} : 1, \nu) + (\boldsymbol{\gamma} : 1, \omega)}.$$

8.5 INVERTED BETA DISTRIBUTION

The beta variate of the second kind, also known as the inverted beta or beta prime variate with parameters ν and ω, denoted $\boldsymbol{I\beta} : \nu, \omega$, is related to the $\boldsymbol{\beta} : \nu, \omega$ variate by

$$\boldsymbol{I\beta} : \nu, \omega \sim (\boldsymbol{\beta} : \nu, \omega)/[1 - (\boldsymbol{\beta} : \nu, \omega)]$$

and to independent standard gamma variates by

$$\boldsymbol{I\beta} : \nu, \omega \sim (\boldsymbol{\gamma} : 1, \nu)/(\boldsymbol{\gamma} : 1, \omega).$$

The inverted beta variate with shape parameters $\nu/2, \omega/2$ is related to the $\boldsymbol{F} : \nu, \ \omega$ variate by

$$\boldsymbol{I\beta} : \nu/2, \omega/2 \sim (\nu/\omega)\boldsymbol{F} : \nu, \ \omega.$$

The pdf is $x^{\nu-1}/[B(\nu, \omega)(1 + x)^{\nu+\omega}]$, $x > 0$.

8.6 NONCENTRAL BETA DISTRIBUTION

The noncentral beta variate $\boldsymbol{\beta} : \nu, \omega, \delta$ is related to the independent noncentral chi-squared in variate $\chi^2 : \nu, \delta$ and the central chi-squared variate $\chi^2 : \omega$ by

$$\frac{\chi^2 : \nu, \delta}{(\chi^2 : \nu, \delta) + (\chi^2 : \omega)} \sim \boldsymbol{\beta} : \nu, \omega, \delta.$$

8.7 BETA BINOMIAL DISTRIBUTION

If the parameter p of a binomial variate $\boldsymbol{B}$: n, p is itself a beta variate $\boldsymbol{\beta} : \nu, \omega$, the resulting variate is a beta binomial variate with probability function

$$\binom{n}{x} \frac{\mathrm{B}(\nu + x, n + \omega - x)}{\mathrm{B}(\nu, \omega)}$$

with mean $n\nu/(\nu + \omega)$ and variance

$$n\nu\omega(n + \nu + \omega)/[(\nu + \omega)^2(1 + \nu + \omega)].$$

This is also called the binomial beta or compound binomial distribution. For integer ν and ω, this corresponds to the negative hypergeometric distribution. For $\nu = \omega = 1$, it corresponds to the discrete rectangular distribution. A multivariate extension of this is the Dirichlet multinomial distribution.

Chapter 9

Binomial Distribution

Applications include the following:

- Estimation of probabilities of outcomes in any set of success or failure trials.
- Estimation of probabilities of outcomes in games of chance.
- Sampling for attributes.

Variate $\boldsymbol{B}: n,\ p$.

Quantile x, number of successes.

Range $0 \leq x \leq n$, x an integer.

The binomial variate $\boldsymbol{B}: n,\ p$ is the number of successes in n-independent Bernoulli trials, where the probability of success at each trial is p and the probability of failure is $q = 1 - p$.

Parameters: n, the Bernoulli trial parameter, a positive integer p, the Bernoulli probability parameter, $0 < p < 1$.

Distribution function	$\sum_{i=0}^{x} \binom{n}{i} p^i q^{n-i}$
Probability function	$\binom{n}{x} p^x q^{n-x}$
Moment generating function	$[p\exp(t) + q]^n$
Probability generating function	$(pt + q)^n$
Characteristic function	$[p\exp(it) + q]^n$
Moments about the origin	
Mean	np
Second	$np(np + q)$
Third	$np[(n-1)(n-2)p^2 + 3p(n-1) + 1]$

Statistical Distributions, Fourth Edition, by Catherine Forbes, Merran Evans, Nicholas Hastings, and Brian Peacock

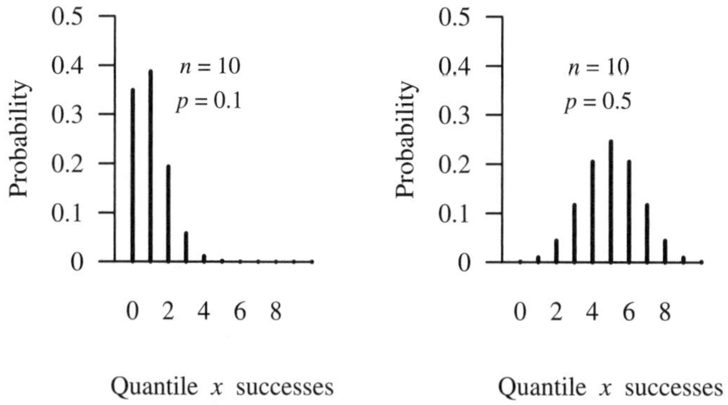

Figure 9.1. Probability function for the binomial variate ***B***: *n*, *p*.

Moments about the mean	
Variance	npq
Third	$npq(q - p)$
Fourth	$np[1 + 3pq(n - 2)]$
Mode	$p(n + 1) - 1 \le x \le p(n + 1)$
Coefficient of skewness	$(q - p)/(npq)^{1/2}$
Coefficient of kurtosis	$3 - \frac{6}{n} + \frac{1}{npq}$
Factorial moments about the mean	
Second	npq
Third	$-2npq(1 + p)$
Coefficient of variation	$(q/np)^{1/2}$

The probability functions for the binomial ***B***: 10, 0.1 and for the ***B***: 10, 0.5 variates are shown in the two panels of Figure 9.1, with the corresponding distribution functions shown in Figure 9.2.

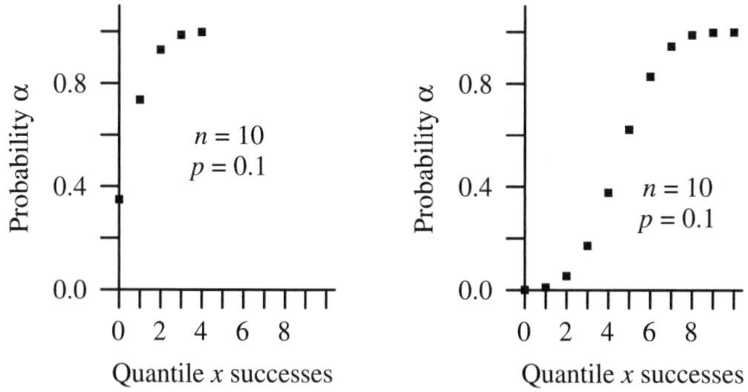

Figure 9.2. Distribution function for the binomial variate ***B***: *n*, *p*.

9.1 VARIATE RELATIONSHIPS

1. For the distribution functions of the binomial variates $\boldsymbol{B}$: n, p and $\boldsymbol{B}$: $n, 1-p$,

$$F_{\boldsymbol{B}}(x : n, p) = 1 - F_{\boldsymbol{B}}(n - x - 1; n, 1 - p).$$

2. The binomial variate $\boldsymbol{B}$: n, p can be approximated by the normal variate with mean np and standard deviation $(npq)^{1/2}$, provided $npq > 5$ and $0.1 \leq p \leq 0.9$ or if $\mathrm{Min}(np, nq) > 10$. For $npq > 25$ this approximation holds for any p.

3. The binomial variate $\boldsymbol{B}$: n, p can be approximated by the Poisson variate with mean np provided $p < 0.1$ and $np < 10$.

4. The binomial variate $\boldsymbol{B}$: n, p with quantile x and the beta variate with shape parameters $x, n - x + 1$ and quantile p are related by

$$\Pr\big[(\boldsymbol{B} : n, p) \geq x\big] = \Pr\big[(\boldsymbol{\beta} : x, n - x + 1) \leq p\big]$$

5. The binomial variate $\boldsymbol{B}$: n, p with quantile x and the $\boldsymbol{F}$ variate with degrees of freedom $2(x + 1), 2(n - x)$, denoted $\boldsymbol{F}$: $(2(x + 1), 2(n - x)$, are related by

$$\begin{aligned}\Pr\big[(\boldsymbol{B} : n, p) \leq x\big] &= 1 - \Pr\big[\big(\boldsymbol{F} : 2(x + 1), 2(n - x)\big)\\ &< p(n - x)/\big[(1 + x)(1 - p)\big]\big].\end{aligned}$$

6. The sum of k-independent binomial variates $\boldsymbol{B}$: $n_i, p; i = 1, \ldots, k$, is the binomial variate $\boldsymbol{B}$: n', p, where

$$\sum_{i=1}^{k}(\boldsymbol{B} : n_i, p) \sim \boldsymbol{B} : n', p, \quad \text{where} \quad n' = \sum_{i=1}^{k} n_i.$$

7. The Bernoulli variate corresponds to the binomial variate with $n = 1$. The sum of n-independent Bernoulli variates $\boldsymbol{B}$: $1, p$ is the binomial variate $\boldsymbol{B}$: n, p.

8. The hypergeometric variate $\boldsymbol{H}$: N, X, n tends to the binomial variate $\boldsymbol{B}$: n, p as N and X tend to infinity and X/N tends to p.

9. The binomial variate $\boldsymbol{B}$: n, p and the negative binomial variate $\boldsymbol{NB}$: x, p (with integer x, which is the Pascal variate) are related by

$$\Pr[(\boldsymbol{B} : n, p) \leq x] = \Pr[(\boldsymbol{NB} : x, p) \geq (n - x)]$$

$$F_{\boldsymbol{NB}}(n - x : x, p) = 1 - F_{\boldsymbol{B}}(x - 1 : n, p).$$

10. The multinomial variate is a multivariate generalization of the binomial variate, where the trials have more than two distinct outcomes.

9.2 PARAMETER ESTIMATION

Parameter	Estimator	Method/Properties
Bernoulli probability, p	x/n	Minimum variance unbiased

9.3 RANDOM NUMBER GENERATION

1. *Rejection Technique*. Select n unit rectangular random numbers. The number of these that are less than p is a random number of the binomial variate $\boldsymbol{B}$: n, p.
2. *Geometric Distribution Method*. If p is small, a faster method may be to add together x geometric random numbers until their sum exceeds $n - x$. The number of such geometric random numbers is a binomial random number.

Chapter 10

Cauchy Distribution

The Cauchy distribution is of mathematical interest due to the absence of defined moments.

Variate C: a, b.

Range $-\infty < x < \infty$.

Location parameter a, the median.

Scale parameter $b > 0$.

Distribution function	$\frac{1}{2} + \frac{1}{\pi}\tan^{-1}\left(\frac{x-a}{b}\right)$
Probability density function	$\left\{\pi b\left[1 + \left(\frac{x-a}{b}\right)^2\right]\right\}^{-1}$
Characteristic function	$\exp(iat - \lvert t \rvert b)$
Inverse distribution function (of probability α)	$a + b\left[\tan\pi\left(\alpha - \frac{1}{2}\right)\right]$
Moments	Do not exist
Cumulants	Do not exist
Mode	a
Median	a

10.1 NOTE

The Cauchy distribution is unimodal and symmetric, with much heavier tails than the normal. The probability density function is symmetric about a, with upper and lower quartiles, $a \pm b$.

The probability density functions of the ***C***: *0, b* variate, for selected values of the scale parameter b, are shown in Figure 10.1.

Statistical Distributions, Fourth Edition, by Catherine Forbes, Merran Evans, Nicholas Hastings, and Brian Peacock

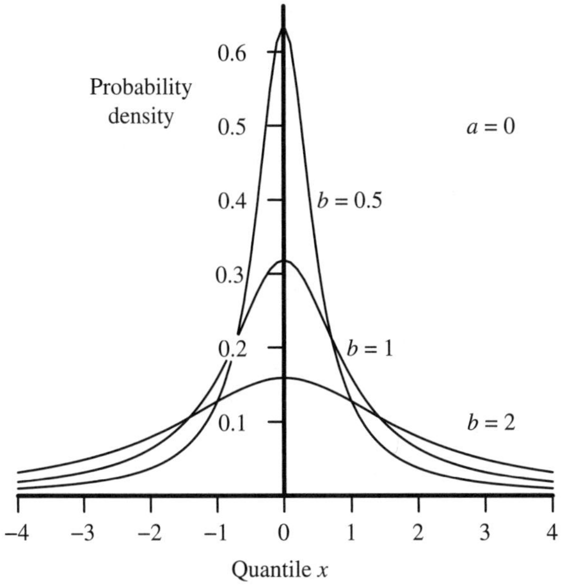

Figure 10.1. Cauchy probability density function C:a,b.

10.2 VARIATE RELATIONSHIPS

The Cauchy variate $\boldsymbol{C}$: a, b is related to the standard Cauchy variate $\boldsymbol{C}$: 0, 1 by

$$\boldsymbol{C} : a, b \sim a + b(\boldsymbol{C} : 0, 1).$$

1. The ratio of two independent unit normal variates $\boldsymbol{N}_1, \boldsymbol{N}_2$ is the standard Cauchy variate $\boldsymbol{C}$: 0, 1.

$$(\boldsymbol{N}_1/\boldsymbol{N}_2) \sim \boldsymbol{C} : 0, 1.$$

2. The standard Cauchy variate is a special case of the Student's $\boldsymbol{t}$ variate with one degree of freedom, $\boldsymbol{t}$: 1.

3. The sum of n-independent Cauchy variates $\boldsymbol{C} : a_i, b_i$ with location parameters $a_i, i = 1, \ldots, n$ and scale parameters $b_i, i = 1, \ldots, n$ is a Cauchy variate $\boldsymbol{C}$: a, b with parameters the sum of those of the individual variates:

$$\sum_{i=1}^{n} (\boldsymbol{C} : a_i, b_i) \sim \boldsymbol{C} : a, b, \quad \text{where} \quad a = \sum_{i=1}^{n} a_i, \quad b = \sum_{i=1}^{n} b_i.$$

The average of n-independent Cauchy variates $\boldsymbol{C}$: a, b is the Cauchy $\boldsymbol{C}$: a, b variate. Hence the distribution is "stable" and infinitely divisible.

4. The reciprocal of a Cauchy variate $\boldsymbol{C}$: a, b is a Cauchy variate $\boldsymbol{C} : a', b'$, where a', b' are given by

$$1/(\boldsymbol{C} : a, b) \sim \boldsymbol{C} : a', b', \text{where} \quad a' = a/(a^2 + b^2), \quad b' = b/(a^2 + b^2).$$

10.3 RANDOM NUMBER GENERATION

The standard Cauchy variate $\boldsymbol{C}$: 0, 1 is generated from the unit rectangular variate $\boldsymbol{R}$ by

$$\boldsymbol{C}: 0, 1 \sim \cot(\pi \boldsymbol{R}) = \tan\left[\pi\left(\boldsymbol{R} - \tfrac{1}{2}\right)\right].$$

10.4 GENERALIZED FORM

Shape parameter $m > 0$, normalizing constant k.
For $m = 1$, this variate corresponds to the Cauchy variate $\boldsymbol{C}$: a, b.
For $a = 0$, this variate corresponds to a Student's $\boldsymbol{t}$ variate with $(2m - 1)$ degrees of freedom, multiplied by $b(2m - 1)^{-1/2}$.

Probability density function	$k\left[1 + \left(\frac{x - a}{b}\right)^2\right]^{-m}, \quad m \geq 1$
	where $k = \Gamma(m)/\left[b\Gamma(1/2)\Gamma\left(m - \frac{1}{2}\right)\right]$
Mean	a
Median	a
Mode	a
rth Moment about the mean	$\begin{cases} \dfrac{\Gamma\left(\frac{r+1}{2}\right)\Gamma\left(m - \frac{r+1}{2}\right)}{\Gamma(1/r)\Gamma(m - 1/r)}, & r \text{ even}, \ r < 2m - 1 \\ 0, & r \text{ odd} \end{cases}$

Chapter 11

Chi-Squared Distribution

Important applications of the chi-squared variate arise from the fact that it is the distribution of the sum of the squares of a number of normal variates. Where a set of data is represented by a theoretical model, the chi-squared distribution can be used to test the goodness of fit between the observed data points and the values predicted by the model, subject to the differences being normally distributed. A particularly common application is the analysis of contingency tables.

Variate $\chi^2 : \nu$.

Range $0 \leq x < \infty$.

Shape parameter ν, degrees of freedom.

Probability density function	$\dfrac{x^{(\nu-2)/2}\exp(-x/2)}{2^{\nu/2}\Gamma(\nu/2)}$ where $\Gamma(\nu/2)$ is the gamma function with argument $\nu/2$
Moment generating function	$(1-2t)^{-\nu/2}, \quad t < \frac{1}{2}$
Laplace transform of the pdf	$(1+2s)^{-\nu/2}, \quad s > -\frac{1}{2}$
Characteristic function	$(1-2it)^{-\nu/2}$
Cumulant generating function	$(-\nu/2)\log(1-2it)$
rth Cumulant	$2^{r-1}\nu(r-1)!, \quad r \geq 1$
rth Moment about the origin	$2^r \prod_{i=0}^{r-1}[\mathrm{i}+(\nu/2)] = \dfrac{2^r\Gamma(r+\nu/2)}{\Gamma(\nu/2)}$
Mean	ν
Variance	2ν
Mode	$\nu - 2, \nu \geq 2$
Median	$\nu - \frac{2}{3}$ (approximately for large ν)

Statistical Distributions, Fourth Edition, by Catherine Forbes, Merran Evans, Nicholas Hastings, and Brian Peacock

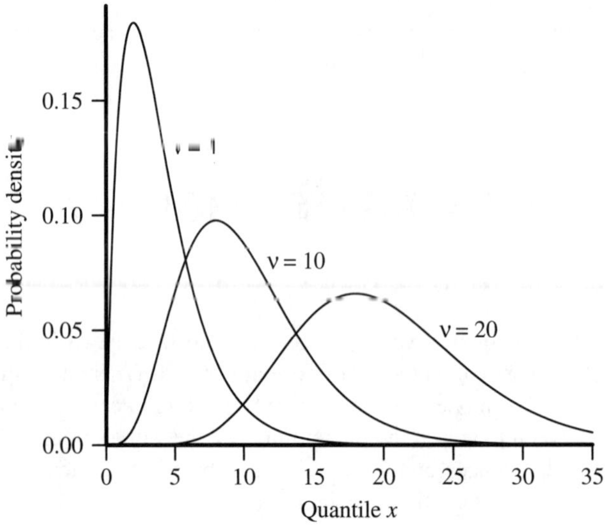

Figure 11.1. Probability density function for the chi-squared variate $\chi^2 : \nu$.

Coefficient of skewness	$2^{3/2}\nu^{-1/2}$
Coefficient of kurtosis	$3 + 12/\nu$
Coefficient of variation	$(2/\nu)^{1/2}$

The probability density function of the $\chi^2 : \nu$ variate is shown in Figure 11.1, with the corresponding distribution function shown in Figure 11.2, for selected values of the degrees of freedom parameter ν.

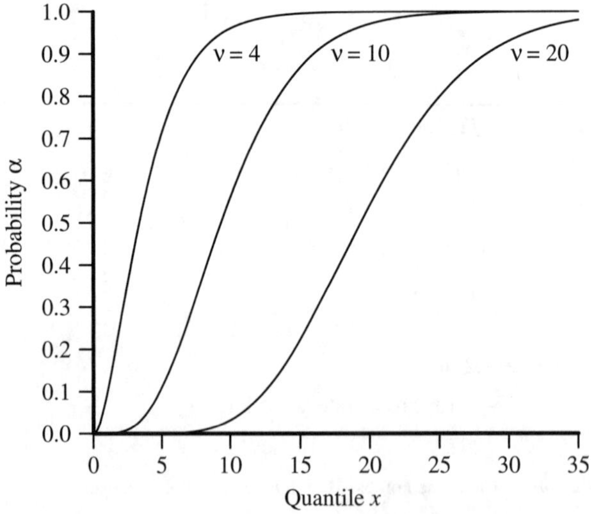

Figure 11.2. Distribution function for the chi-squared variate $\chi^2 : \nu$.

11.1 VARIATE RELATIONSHIPS

1. The chi-squared variate with ν degrees of freedom is equal to the gamma variate with scale parameter 2 and shape parameter $\nu/2$, or equivalently is twice the gamma variate with scale parameter 1 and shape parameter $\nu/2$.

$$\begin{aligned}\chi^2 : \nu &\sim \gamma : 2, \nu/2 \\ &\sim 2(\gamma : 1, \nu/2).\end{aligned}$$

Properties of the gamma variate apply to the chi-squared variate $\chi^2 : \nu$. The chi-squared variate χ^2: 2 is the exponential variate E: 2.

2. The independent chi-squared variates with ν and ω degrees of freedom, denoted $\chi^2 : \nu$ and $\chi^2 : \omega$, respectively, are related to the F variate with degrees of freedom ν, ω, denoted F: ν, ω by

$$F : \nu, \omega \sim \frac{(\chi^2 : \nu)/\nu}{(\chi^2 : \omega)/\omega}.$$

3. As ω tends to infinity, ν times the F variate F: ν, ω tends to the chi-squared variate $\chi^2 : \nu$.

$$\chi^2 : \nu \approx \nu(F : \nu, \omega) \quad \text{as} \quad \omega \to \infty.$$

4. The chi-squared variate $\chi^2 : \nu$ is related to the Student's t variate with ν degrees of freedom, denoted t: ν, and the independent unit normal variate N: 0, 1 by

$$t : \nu \sim \frac{N : 0, 1}{[(\chi^2 : \nu)/\nu]^{1/2}}.$$

5. The chi-squared variate $\chi^2 : \nu$ is related to the Poisson variate with mean $x/2$, denoted P: $x/2$, by

$$\Pr[(\chi^2 : \nu) > x] = \Pr[(P : x/2) \le ((\nu/2) - 1)].$$

Equivalent statements in terms of the distribution function F and inverse distribution function G are

$$1 - F_{\chi^2}(x : \nu) = F_P([(\nu/2) - 1] : x/2)$$
$$G_{\chi^2}((1-\alpha) : \nu) = x \quad \Leftrightarrow \quad G_P(\alpha : x/2) = (\nu/2) - 1.$$

$0 \le x < \infty$; $\nu/2$ a positive integer; $0 < \alpha < 1$; α denotes probability.

6. The chi-squared variate $\chi^2 : \nu$ is equal to the sum of the squares of ν-independent unit normal variates, N: 0, 1.

$$\chi^2 : \nu \sim \sum_{i=1}^{\nu} (N : 0, 1)_i^2 \sim \sum_{i=1}^{\nu} \left(\frac{(N : \mu_i, \sigma_i) - \mu_i}{\sigma_i} \right)^2$$

7. The sum of independent chi-squared variates is also a chi-squared variate:

$$\sum_{i-1}^{n}(\chi^2 : \nu_i) \sim \chi^2 : \nu, \quad \text{where} \quad \nu = \sum_{i=1}^{n} \nu_i.$$

8. The chi-squared variate $\chi^2 : \nu$ for ν large can be approximated by transformations of the normal variate.

$$\chi^2 : \nu \approx \tfrac{1}{2}[(2\nu - 1)^{1/2} + (N : 0, 1)]^2$$
$$\chi^2 : \nu \approx \nu[1 - 2/(9\nu) + [2/(9\nu)]^{1/2}(N : 0, 1)]^3.$$

The first approximation of Fisher is less accurate than the second of Wilson–Hilferty.

9. Given n normal variates N: μ, σ, the sum of the squares of their deviations from their mean is the variate $\sigma^2\chi^2 : n - 1$. Define variates $\bar{x}, s^2$ as follows:

$$\bar{x} \sim \frac{1}{n}\sum_{i=1}^{n}(N : \mu, \sigma)_i, \quad s^2 \sim \frac{1}{n}\sum_{i=1}^{n}\left[(N : \mu, \sigma)_i - \bar{x}\right]^2.$$

Then $ns^2/\sigma^2 \sim \chi^2 : n - 1$.

10. Consider a set of n_1-independent normal variates N: μ_1, σ, and a set of n_2-independent normal variates N: μ_2, σ (note same σ) and define variates $\bar{x}_1, \bar{x}_2, s_1^2, s_2^2$ as follows:

$$\bar{x}_1 \sim \frac{1}{n_1}\sum_{i=1}^{n_1}(N : \mu_1, \sigma)_i; \quad s_1^2 \sim \frac{1}{n_1}\sum_{i=1}^{n_1}\left[(N : \mu_1, \sigma)_i - \bar{x}_1\right]^2$$

$$\bar{x}_2 \sim \frac{1}{n_2}\sum_{j=1}^{n_2}(N : \mu_2, \sigma)_j; \quad s_2^2 \sim \frac{1}{n_2}\sum_{j=1}^{n_2}\left[(N : \mu_2, \sigma)_j - \bar{x}_2\right]^2.$$

Then

$$\left(n_1 s_1^2 + n_2 s_2^2\right)/\sigma^2 \sim \chi^2 : n_1 + n_2 - 2.$$

11.2 RANDOM NUMBER GENERATION

For independent N: 0, 1 variates

$$\chi^2 : \nu \sim \sum_{i=1}^{\nu}(N : 0, 1)_i^2.$$

See also the gamma distribution.

11.3 CHI DISTRIBUTION

The positive square root of a chi-squared variate, $\chi^2 : \nu$, has a chi distribution with shape parameter ν, the degrees of freedom. The probability density function is

$$x^{\nu-1}\exp(-x^2/2)/\left[2^{\nu/2-1}\Gamma(\nu/2)\right]$$

and the rth central moment about the origin is

$$2^{r/2}\Gamma\left[(\nu+r)/2\right]\Gamma(\nu/2)$$

and the mode is $\sqrt{\nu-1}$, $\nu \geq 1$.

This chi variate, $\chi : \nu$, corresponds to the Rayleigh variate for $\nu = 2$ and the Maxwell variate with unit scale parameter for $\nu = 3$. Also, $|N : 0, 1| \sim \chi : 1$.

Chapter 12

Chi-Squared (Noncentral) Distribution

An application of the noncentral chi-squared distribution is to describe the size of cohorts of wildlife species in predefined unit-area blocks. In some cases the whole cohort will be found in a particular block, whereas in others some of the cohort may have strayed outside the predefined area.

The chi-squared (noncentral) distribution is also known as the generalized Rayleigh, Rayleigh–Rice, or Rice distribution.

Variate $\chi^2 : \nu, \delta$.

Range $0 < x < \infty$.

Shape parameters $\nu > 0$, the degrees of freedom, and $\delta \geq 0$, the noncentrality parameter.

Probability density function	$\dfrac{\exp[-\frac{1}{2}(x+\delta)]}{2^{\nu/2}} \sum_{j=0}^{\infty} \dfrac{x^{\nu/2+j-1}\delta^j}{\Gamma(\nu/2+j)2^{2j}j!}$
Moment generating function	$(1-2t)^{-\nu/2}\exp[\delta t/(1-2t)],\ t < \frac{1}{2}$
Characteristic function	$(1-2it)^{-\nu/2}\exp[\delta it/(1-2it)]$
Cumulant generating function	$-\frac{1}{2}\nu\log(1-2it) + \delta it/(1-2it)$
rth Cumulant	$2^{r-1}(r-1)!(\nu + r\delta)$
rth Moment about the origin	$2^r\Gamma\left(r+\frac{\nu}{2}\right)\sum_{j=0}^{r}\binom{r}{j}\left(\frac{\delta}{2}\right)^j \Big/ \Gamma\left(j+\frac{\nu}{2}\right)$
Mean	$\nu + \delta$
Moments about the mean	
Variance	$2(\nu + 2\delta)$
Third	$8(\nu + 3\delta)$
Fourth	$48(\nu + 4\delta) + 12(\nu + 2\delta)^2$

Statistical Distributions, Fourth Edition, by Catherine Forbes, Merran Evans, Nicholas Hastings, and Brian Peacock

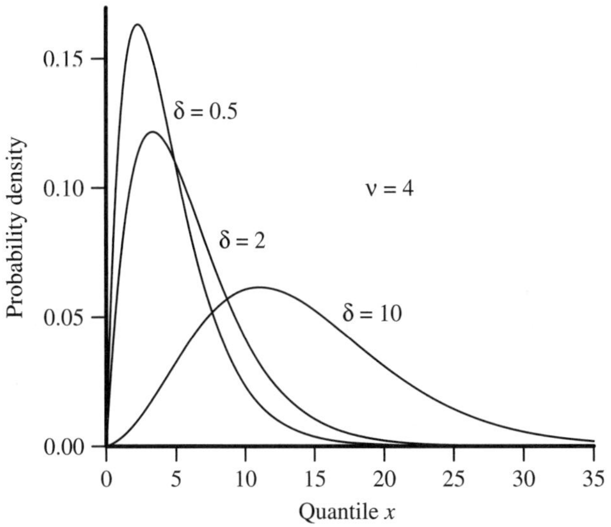

Figure 12.1. Probability density function for the (noncentral) chi-squared variate $\chi^2 : \nu, \ \delta$.

Coefficient of skewness	$8^{1/2}(\nu + 3\delta)/(\nu + 2\delta)^{3/2}$
Coefficient of kurtosis	$3 + 12(\nu + 4\delta)/(\nu + 2\delta)^2$
Coefficient of variation	$[2(\nu + 2\delta)]^{1/2}/(\nu + \delta)$

The probability density function of the $\chi^2 : 4, \delta$ variate is shown in Figure 12.1 for selected values of the noncentrality parameter δ.

12.1 VARIATE RELATIONSHIPS

1. Given ν-independent standard normal variates N: 0, 1, then, the noncentral chi-squared variate corresponds to

$$\chi^2 : \nu, \delta \sim \sum_{i=1}^{\nu} [(N : 0, 1)_i + \delta_i]^2 \sim \sum_{i=1}^{\nu} (N : \delta_i, 1)^2,$$

where $\delta = \sum_{i=1}^{\nu} \delta_i^2$.

2. The sum of n-independent noncentral chi-squared variates $\chi^2 : \nu_i, \delta_i, i = 1, \ldots, n$, is a noncentral chi-squared variate $\chi^2 : \nu, \delta$.

$$\chi^2 : \nu, \delta \sim \sum_{i=1}^{n} (\chi^2 : \nu_i, \delta_i), \quad \text{where} \quad \nu = \sum_{i=1}^{n} \nu_i, \delta = \sum_{i=1}^{n} \delta_i.$$

3. The noncentral chi-squared variate $\chi^2 : \nu, \delta$ with zero noncentrality parameter $\delta = 0$ is the (central) chi-squared variate $\chi^2 : \nu$.

4. The standardized noncentral chi-squared variate $\chi^2 : \nu, \delta$ tends to the standard normal variate N: 0, 1, either when ν tends to infinity as δ remains fixed or when δ tends to infinity as ν remains fixed.

5. The noncentral chi-squared variate $\chi^2 : \nu, \delta$ (for ν even) is related to two independent Poisson variates with parameters $\nu/2$ and $\delta/2$, denoted P: $\nu/2$ and P: $\delta/2$, respectively, by

$$\Pr[(\chi^2 : \nu, \delta) \leq x] = \Pr[[(P : \nu/2) - (P : \delta/2)] \geq \nu/2].$$

6. The independent noncentral chi-squared variate $\chi^2 : \nu, \delta$ and central chi-squared variate $\chi^2 : \omega$ are related to the noncentral F variate F: ν, ω, δ by

$$F : \nu, \omega, \delta \sim \frac{(\chi^2 : \nu, \delta)/\nu}{(\chi^2 : \omega)/\omega}.$$

Chapter 13

Dirichlet Distribution

The standard or Type I Dirichlet is a multivariate generalization of the beta distribution.

Vector quantile with elements $x_1, \ldots, x_k$.

Range $x_i \geq 0$, $\sum_{i=1}^{k} x_i \leq 1$.

Parameters $c_i > 0$, $i = 1, \ldots, k$ and c_0.

Probability density function $\dfrac{\Gamma\left(\sum\limits_{i=0}^{k} c_i\right)}{\prod\limits_{i=0}^{k} \Gamma(c_i)} \prod\limits_{i=1}^{k} x_i^{c_i - 1} \left(1 - \sum\limits_{i=1}^{k} x_i\right)^{c_0 - 1}$.

For individual elements (with $c = \sum_{i=0}^{k} c_i$):

Mean	c_i/c
Variance	$c_i(c - c_i)/[c^2(c+1)]$
Covariance	$-c_i c_j/[c^2(c+1)]$

13.1 VARIATE RELATIONSHIPS

1. The elements X_i, $i = 1, \ldots, k$, of the Dirichlet multivariate vector are related to independent standard gamma variates with shape parameters c_i, $i = 0, \ldots, k$, by

$$X_i \sim \frac{\gamma : 1, c_i}{\left(\sum\limits_{j=0}^{k} (\gamma : 1, c_j)\right)}, \quad i = 1, \ldots, k$$

Statistical Distributions, Fourth Edition, by Catherine Forbes, Merran Evans, Nicholas Hastings, and Brian Peacock

and independent chi-squared variates with shape parameters $2\nu_i, i = 0, \ldots, k$, by

$$X_i \sim \frac{\chi^2 : 2\nu_i}{\sum_{j=0}^{k} (\chi^2 : 2\nu_j)}, \quad i = 1, \ldots, k.$$

2. For $k = 1$, the Dirichlet univariate is the beta variate $\boldsymbol{\beta} : \nu, \omega$ with parameters $\nu = c_1$ and $\omega = c_0$. The Dirichlet variate can be regarded as a multivariate generalization of the beta variate.
3. The marginal distribution of X_i is the standard beta distribution with parameters

$$\nu = c_i \quad \text{and} \quad \omega = \sum_{j=0}^{k} c_j - c_i.$$

4. The Dirichlet variate with parameters np_i is an approximation to the multinominal variate, for np_i not too small for every i.

13.2 DIRICHLET MULTINOMIAL DISTRIBUTION

The Dirichlet multinominal distribution is the multivariate generalization of the beta binomial distribution. It is also known as the compound multinomial distribution and, for integer parameters, the multivariate negative hypergeometric distribution.

It arises if the parameters $p_i, i = 1, \ldots, k$, of the multinomial distribution follow a Dirichlet distribution. It has probability function

$$\frac{n!\Gamma\left(\sum_{j=1}^{k} c_j\right)}{\Gamma\left(n + \sum_{j=1}^{k} c_j\right)} \prod_{j=1}^{k} \frac{x_j + c_j}{c_j}, \quad \sum_{i=1}^{k} x_i = n, \quad x_i \geq 0.$$

The mean of the individual elements x_i is nc_i/c, where $c = \sum_{j=1}^{k} c_j$, and the variances and covariances correspond to those of a multinomial distribution with $p_i = c_i/c$. The marginal distribution of X_i is a beta binomial.

Chapter 14

Empirical Distribution Function

An empirical distribution function is one that is estimated directly from sample data, without assuming an underlying algebraic form of the distribution model.

Variate X.

Quantile x.

Distribution function (unknown) $F(x)$.

Empirical distribution function (edf) $F_E(x)$.

14.1 ESTIMATION FROM UNCENSORED DATA

Consider a data sample from X, of size n, with observations x_i, $i = 1, n$, arranged in nondecreasing order. Here i is referred to as the order-number of observation i. Estimates are made of the value of the distribution function $F(x)$, at points corresponding to the observed quantile values, x_i.

Estimates of the empirical distribution function (edf) $F_E(x_i)$ at x_i are the following:

Kaplan– Meier estimate	i/n
Mean rank estimate	$i/(n+1)$
Median rank estimate	$(i-0.3)/(n+0.4)$

14.2 ESTIMATION FROM CENSORED DATA

The data consist of a series of *events*, some of which are *observations* and some are random *censorings*. Let the events be arranged in nondecreasing order of their quantile value. An observation order-number i, a censoring order-number j, an event

Statistical Distributions, Fourth Edition, by Catherine Forbes, Merran Evans, Nicholas Hastings, and Brian Peacock

order-number e, and a modified (observation) order-number m_i are used. A modified order-number is an adjusted observation order-number that allows for the censored items. The modified order-numbers and then the distribution function value at quantile x_i are calculated with the modified order-number m_i replacing the observation order-number i in the median rank equation above.

i	= observation order-number (excludes censorings)
I	= total number of observations (excludes censorings)
e	= event order-number (observations and censorings)
n	= total number of events (observations and censorings)
j	= censoring order-number
e_i	= event-number of observation i
e_j	= event-number of censoring j
m_i	= modified order-number of observation i
$C(i)$	= set of censoring occurring at or after observation $i-1$ and before observation i (this set may be empty)
x_i	= the quantile value (e.g., age at failure) for observation i
x_e	= the quantile value for event e, which may be an observation or a censoring
x_j	= the quantile value for censoring j
m_i^*	$= n+1-m_i$
e_i^*	$= n+1-e_i$
α_j	= the proportion of the current interobservation interval that has elapsed when censoring j occurs
x_0	$= m_0 = e_0 = 0$

For censoring j in the set $C(i)$, α_j is defined by

$$\alpha_j = (x_j - x_{i-1})/(x_i - x_{i-1}).$$

α_j is the proportion of the interval between observation $i-1$ and observation i, which elapses before censoring j occurs. The method used is described in Bartlett and Hastings (1998). The Herd-Johnson method described by d'Agostino and Stephens (1986) is equivalent to assuming that all values of α_j are zero.

The formula for the modified order-number is

$$m_i - m_{i-1} = m_{i-1}^* \left(1 - \frac{e_i^*}{e_{i-1}^*} \prod_{C(i)} \frac{e_j^* + 1 - \alpha_j}{e_j^* - \alpha_j}\right).$$

Here, the product is taken over suspensions in the set $C(i)$. If this set is empty the product term has value 1, and

$$m_i - m_{i-1} = m_{i-1}^*/e_{i-1}^*$$

$$F_E(x_i) = (m_i - 0.3)/(n + 0.4).$$

Table 14.1: Modified Order-Numbers and Median Ranks

Event order-number e	Hours run, x	Status	Observation order-number i	Modified order-number m_i	Median rank, $F(x)$ $(m_i - 0.3)/(n + 0.4)$
1	3,895	Failure	1	1	0.1591
2	4,733	Failure	2	2	0.3864
3	7,886	Censoring			
4	9,063	Failure	3	3.2137	0.6622

14.3 PARAMETER ESTIMATION

$$\hat{\mu} = \left(\frac{1}{I}\right) \sum_{e=1}^{N} x_e$$

14.4 EXAMPLE

An example of reliability data, relating to the lives of certain mechanical components, is shown in the first three columns of Table 14.1. The observations are ages (in kilometers run) at failure and the censoring is the age of an item that has run but not failed.

To estimate the empirical distribution function, first calculate the modified order-numbers using the equations in Section 14.2. For events prior to the first censoring, $m_1 = 1, m_2 = 2$ is obtained. Event 3 is a censoring.

$$\alpha_3 = (7886 - 4733)/(9063 - 4733) = 0.7282$$

$$m_3 - m_2 = 3\left[1 - \frac{1}{3} \times (3 - 0.7282)\right] \Big/ (2 - 0.7282) = 1.2137$$

$$m_3 = 3.2137$$

Median ranks are calculated as in Section 14.1. The results are summarized in Table 14.1 and shown in graphical form in Figure 14.1.

14.5 GRAPHICAL METHOD FOR THE MODIFIED ORDER-NUMBERS

A square grid (Figure 14.2) is drawn with the sides made up of $n + 1$ small squares, where n is the total number of events. The bottom edge of the grid is numbered with observation order-numbers $i = 1, \ldots, I$. The top edge of the grid is numbered with censoring order-numbers, starting from the top right-hand corner and working to the left. The left-hand side of the grid represents the scale of modified order-numbers.

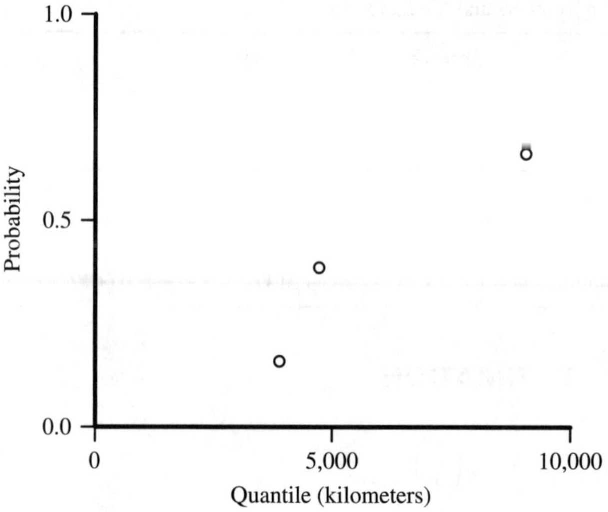

Figure 14.1. Estimated empirical distribution function under censoring.

If there are no censorings, the observation order-numbers and the modified order-numbers are the same. This situation would be represented by a 45° diagonal line across the large square. In Figure 14.2, the diagonal line starts out at 45°, as initially there are no censorings, and for the first two observations the observation order-number and the modified order-number are the same. When there is a censoring, *j*,

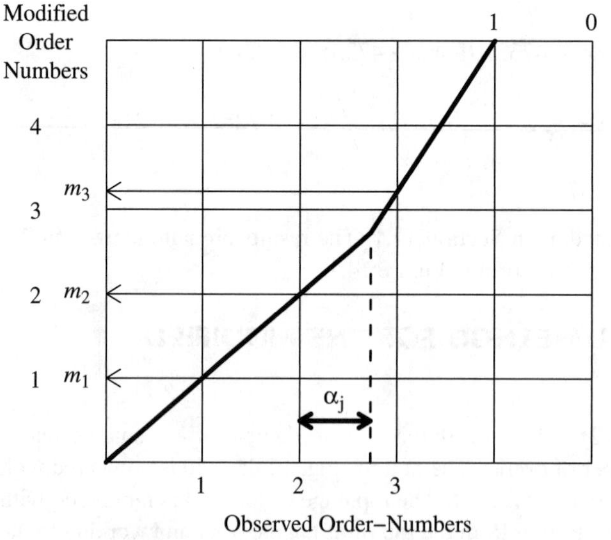

Figure 14.2. Graphical method for the modified order-numbers.

between observation $i-1$ and observation i, the censoring is indicated by a dotted vertical line placed a proportion α_j between observations. The vertical dotted line i between observation 2 and observation 3 illustrates this. The gradient of the diagonal line increases at each censoring, becoming directed toward the corresponding censoring order-number on the top scale of the square. In the example, there is only one censoring, so the gradient of the diagonal line increases only once. The modified order-numbers are read off from the left-hand vertical scale, by drawing horizontal lines across from the points where the vertical lines through the observation numbers intersect the diagonal line.

14.6 MODEL ACCURACY

For any given distribution model, let $F_M(x_i)$ be the distribution function value at quantile x_i, where the edf value is $F_E(x_i)$. Let q_i^2 be the square of the difference between the model value $F_M(x_i)$ and the edf value $F_E(x_i)$.

$$q_i^2 = [F_M(x_i) - F_E(x_i)]^2.$$

The mean square error between the edf points and the model distribution function is given by the Cramer–von Mises statistic:

$$Q^2 = \left(\frac{1}{I}\right)\sum_{i=1}^{I} q_i^2.$$

A, the model accuracy, is defined (in percentage terms) by

$$A = 100(1 - Q).$$

If the edf points all lie exactly on the model distribution function curve, the model accuracy is 100%.

Chapter 15

Erlang Distribution

The Erlang variate is the sum of a number of exponential variates. It was developed as the distribution of waiting time and message length in telephone traffic. If the durations of individual calls are exponentially distributed, the duration of a succession of calls has an Erlang distribution.

The Erlang variate is a gamma variate with shape parameters c, an integer. The diagrams, notes on parameter estimation, and variate relationships for the gamma variate apply to the Erlang variate.

Variate $\boldsymbol{\gamma}$: b, c.

Range $0 \le x < \infty$.

Scale parameter $b > 0$. Alternative parameter $\lambda = 1/b$.

Shape parameter $c > 0$, c an integer for the Erlang distribution.

Distribution function	$1 - \left[\exp\left(-\frac{x}{b}\right)\right]\left(\sum_{i=0}^{c-1}\frac{(x/b)^i}{i!}\right)$
Probability density function	$\frac{(x/b)^{c-1}\exp(-x/b)}{b(c-1)!}$
Survival function	$\exp\left(-\frac{x}{b}\right)\left(\sum_{i=0}^{c-1}\frac{(x/b)^i}{i!}\right)$
Hazard function	$\frac{(x/b)^{c-1}}{b(c-1)!\sum_{i=0}^{c-1}\frac{(x/b)^i}{i!}}$
Moment generating function	$(1-bt)^{-c}, t < 1/b$
Laplace transform of the pdf	$(1+bs)^{-c}$
Characteristic function	$(1-ibt)^{-c}$
Cumulant generating function	$-c\log(1-ibt)$

Statistical Distributions, Fourth Edition, by Catherine Forbes, Merran Evans, Nicholas Hastings, and Brian Peacock

rth Cumulant	$(r-1)!cb^r$
rth Moment about the origin	$b^r \prod_{i=0}^{r-1}(c+i)$
Mean	bc
Variance	b^2c
Mode	$b(c-1),\ c\geq 1$
Coefficient of skewness	$2c^{-1/2}$
Coefficient of kurtosis	$3+6/c$
Coefficient of variation	$c^{-1/2}$

15.1 VARIATE RELATIONSHIPS

1. If $c = 1$, the Erlang reduces to the exponential distribution $\boldsymbol{E}$: b.
2. The Erlang variate with scale parameter b and shape parameter c, denoted $\boldsymbol{\gamma}$: b, c, is equal to the sum of c-independent exponential variates with mean b, denoted $\boldsymbol{E}$: b.

$$\boldsymbol{\gamma} : b, c \sim \sum_{i=1}^{c} (\boldsymbol{E} : b)_i, \quad c \text{ a positive integer}$$

3. For other properties see the gamma distribution.

15.2 PARAMETER ESTIMATION

See gamma distribution.

15.3 RANDOM NUMBER GENERATION

$$\boldsymbol{\gamma} : b, c \sim -b \log \left(\prod_{i=1}^{c} \boldsymbol{R}_i \right)$$

where $\boldsymbol{R}_i$ are independent rectangular unit variates.

Chapter 16

Error Distribution

The error distribution is also known as the exponential power distribution or the general error distribution.

Range $-\infty < x < \infty$.

Location parameter $-\infty < a < \infty$, the mean.

Scale parameter $b > 0$.

Shape parameter $c > 0$. Alternative parameter $\lambda = 2/c$.

Probability density function	$\dfrac{\exp[-(\lvert x-a\rvert/b)^{2/c}/2]}{b(2^{c/2+1})\Gamma(1+c/2)}$
Mean	a
Median	a
Mode	a
rth Moment about the mean	$\begin{cases} b^r 2^{rc/2}\dfrac{\Gamma((r+1)c/2)}{\Gamma(c/2)}, & r \text{ even} \\ 0, & r \text{ odd} \end{cases}$
Variance	$\dfrac{2^c b^2 \Gamma(3c/2)}{\Gamma(c/2)}$
Mean deviation	$\dfrac{2^{c/2} b\Gamma(c)}{\Gamma(c/2)}$
Coefficient of skewness	0
Coefficient of kurtosis	$\dfrac{\Gamma(5c/2)\Gamma(c/2)}{[\Gamma(3c/2)]^2}$

Statistical Distributions, Fourth Edition, by Catherine Forbes, Merran Evans, Nicholas Hastings, and Brian Peacock

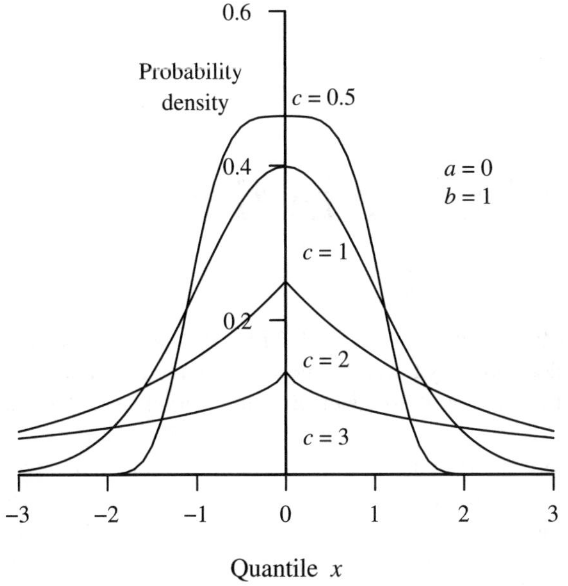

Figure 16.1. Probability density function for the error variate.

16.1 NOTE

Distributions are symmetric, and for $c > 1$ are leptokurtic and for $c < 1$ are platykurtic.

The probability density function of the error distribution variate is shown in Figure 16.1, with location parameter $a = 0$, scale parameter $b = 1$ and for selected values of the shape parameter, c.

16.2 VARIATE RELATIONSHIPS

1. The error variate with $a = 0, b = c = 1$ corresponds to a standard normal variate $\boldsymbol{N}$: 0, 1.
2. The error variate with $a = 0, b = \frac{1}{2}, c = 2$ corresponds to a Laplace variate $\boldsymbol{L}$: 0, 1.
3. As c tends to zero, the error variate tends to a rectangular variate with range $(a - b, a + b)$.

Chapter 17

Exponential Distribution

This is a distribution of the time to an event when the probability of the event occurring in the next small time interval does not vary through time. It is also the distribution of the time between events when the number of events in any time interval has a Poisson distribution.

The exponential distribution has many applications. Examples include the time to decay of a radioactive atom and the time to failure of components with constant failure rates. It is used in the theory of waiting lines or queues, which are found in many situations: from the gates at the entrance to toll roads through the time taken for an answer to a telephone enquiry, to the time taken for an ambulance to arrive at the scene of an accident. For exponentially distributed times, there will be many short times, fewer longer times, and occasional very long times. See Chapter 38.

The exponential distribution is also known as the negative exponential distribution.

Variate $\boldsymbol{E}$: b.

Range $0 \leq x < +\infty$.

Scale parameter $b > 0$, the mean.

Alternative parameter λ, the hazard function (hazard rate), $\lambda = 1/b$.

Distribution function	$1 - \exp(-x/b)$
Probability density function	$(1/b)\exp(-x/b) = \lambda\exp(-\lambda x)$
Inverse distribution function (of probability α)	$b\log[1/(1-\alpha)] = -b\log(1-\alpha)$
Survival function	$\exp(-x/b)$
Inverse survival function (of probability α)	$b\log(1/\alpha) = -b\log(\alpha)$
Hazard function	$1/b = \lambda$

Statistical Distributions, Fourth Edition, by Catherine Forbes, Merran Evans, Nicholas Hastings, and Brian Peacock

Cumulative hazard function	x/b
Moment generating function	$1/(1-bt),\ t < 1/b = \lambda/(\lambda - t)$
Laplace transform of the pdf	$1/(1+bs), \quad s > -1/b$
Characteristic function	$1/(1-ibt)$
Cumulant generating function	$-\log(1-ibt)$
rth Cumulant	$(r-1)!b^r, \quad r \geq 1$
rth Moment about the origin	$r!b^r$
Mean	b
Variance	b^2
Mean deviation	$2b/e$, where e is the base of natural logarithms
Mode	0
Median	$b \log 2$
Coefficient of skewness	2
Coefficient of kurtosis	9
Coefficient of variation	1
Information content	$\log_2(eb)$

17.1 NOTE

The exponential distribution is the only continuous distribution characterized by a "lack of memory." An exponential distribution truncated from below has the same

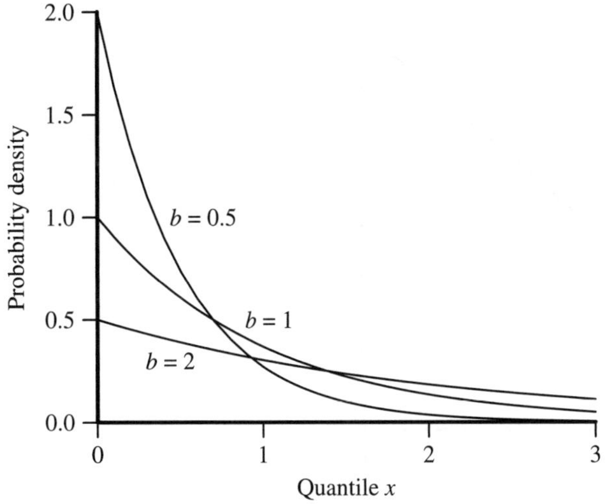

Figure 17.1. Probability density function for the exponential variate $\boldsymbol{E}: b$.

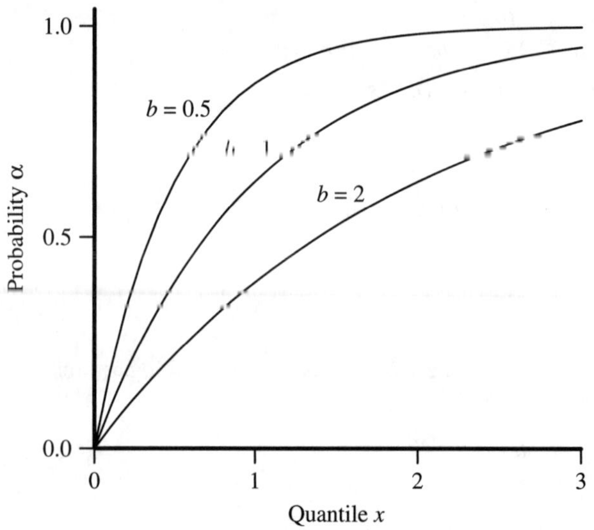

Figure 17.2. Distribution function for the exponential variate $\boldsymbol{E}:b$.

distribution with the same parameter b. The geometric distribution is its discrete analogue. The hazard rate is constant.

The probability density function of the $\boldsymbol{E}:b$ variate is shown in Figure 17.1 for selected values of the scale parameter, b. The corresponding distribution function and cumulative hazard function are shown in Figures 17.2 and 17.3, respectively.

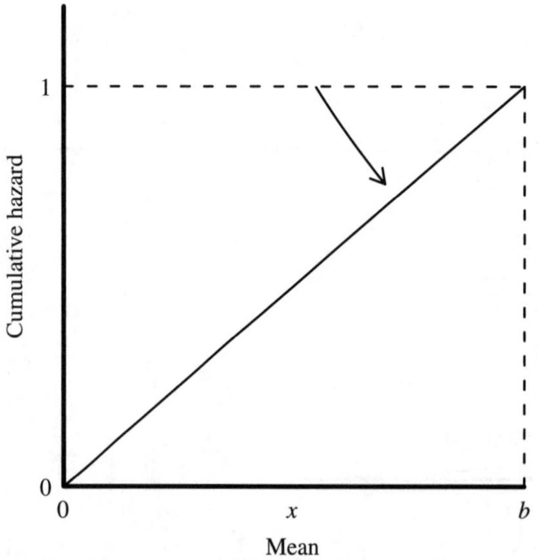

Figure 17.3. Cumulative hazard function for the exponential variate $\boldsymbol{E}:b$.

17.2 VARIATE RELATIONSHIPS

$(\mathbf{E}: b)/b \sim \boldsymbol{E}: 1$, the unit exponential variate.

1. The exponential variate $\boldsymbol{E}$: b is a special case of the gamma variate $\boldsymbol{\gamma}$: b, c corresponding to shape parameter $c = 1$.

$$\boldsymbol{E} : b \sim \boldsymbol{\gamma} : b, 1$$

2. The exponential variate $\boldsymbol{E}$: b is a special case of the Weibull variate $\boldsymbol{W}$: b, c corresponding to shape parameter $c = 1$.

$$\boldsymbol{E} : b \sim \boldsymbol{W} : b, 1$$

$\boldsymbol{E}$: 1 is related to Weibull variate $\boldsymbol{W}$: b, c

$$b(\boldsymbol{E} : 1)^{1/c} \sim \boldsymbol{W} : b,\ c.$$

3. The exponential variate $\mathbf{E}$: b is related to the unit rectangular variate $\boldsymbol{R}$ by

$$\boldsymbol{E} : b \sim -b \log \boldsymbol{R}.$$

4. The sum of c-independent exponential variates, $\boldsymbol{E}$: b, is the Erlang (gamma) variate $\boldsymbol{\gamma}$: b, c, with integer parameter c.

$$\sum_{i=1}^{c} (\boldsymbol{E} : b)_i \sim \boldsymbol{\gamma} : b, c$$

5. The difference of the two independent exponential variates, $(\boldsymbol{E} : b)_1$ and $(\boldsymbol{E} : b)_2$, is the Laplace variate with parameters 0, b, denoted $\boldsymbol{L}$: 0, b.

$$\boldsymbol{L} : 0, b \sim (\boldsymbol{E} : b)_1 - (\boldsymbol{E} : b)_2$$

If $\boldsymbol{L}$: a, b is the Laplace variate, $\boldsymbol{E} : b \sim |(\boldsymbol{L} : a, b) - a|$.

6. The exponential variate $\boldsymbol{E}$: b is related to the standard power function variate with shape parameter c, here denoted $\boldsymbol{X}$: c, for $c = 1/b$

$$\boldsymbol{X} : c \sim \exp(-\boldsymbol{E} : b) \quad \text{for} \quad c = 1/b$$

and the Pareto variate with shape parameter c, here denoted $\boldsymbol{X}$: a, c for $c = 1/b$, by

$$\boldsymbol{X} : a, c \sim a \exp(\boldsymbol{E} : b) \quad \text{for } c = 1/b.$$

7. The exponential variate $\boldsymbol{E}$: b is related to the Gumbel extreme value variate $\boldsymbol{V}$: a,b by

$$\boldsymbol{V} : a, b \sim a - \log(\boldsymbol{E} : b).$$

8. Let $\boldsymbol{Y}$ be a random variate with a continuous distribution function F_Y. Then the standard exponential variate $\boldsymbol{E}$: 1 corresponds to $\boldsymbol{E} : 1 \sim -\log(1 - F_Y)$.

17.3 PARAMETER ESTIMATION

Parameter	Estimator	Method/Properties
b	$\bar{x}$	Unbiased, maximum likelihood

17.4 RANDOM NUMBER GENERATION

Random numbers of the exponential variate $\boldsymbol{E}$: b can be generated from random numbers of the unit rectangular variate $\boldsymbol{R}$ using the relationship

$$\boldsymbol{E} : b \sim -b \log \boldsymbol{R}.$$

Chapter 18

Exponential Family

Variate can be discrete or continuous and uni- or multidimensional.

Parameter θ can be uni- or multidimensional.

The exponential family is characterized by having a pdf or pf of the form

$$\exp\left[A(\theta)\cdot B(x) + C(x) + D(\theta)\right].$$

18.1 MEMBERS OF THE EXPONENTIAL FAMILY

These include the univariate Bernoulli, binomial, Poisson, geometric, gamma, normal, inverse Gaussian, logarithmic, Rayleigh, and von Mises distributions. Multivariate distributions include the multinomial, multivariate normal, Dirichlet, and Wishart.

18.2 UNIVARIATE ONE-PARAMETER EXPONENTIAL FAMILY

The natural exponential family has $B(x) = x$, with $A(\theta)$ the natural or canonical parameter.

For $A(\theta) = \theta$:

Probability (density) function	$\exp\left[\theta x + C(x) + D(\theta)\right]$
Characteristic function	$\exp\left[D(\theta) - D(\theta + it)\right]$
Cumulant generating function	$D(\theta) - D(\theta + it)$
rth Cumulant	$-\dfrac{d^r}{d\theta^r} D(\theta)$

Particular cases are:

Binomial $\boldsymbol{B}$: n, p for $\theta = p$

$$A(\theta) = \log[\theta/(1-\theta)] = \log(p/q), \quad B(x) = x$$

$$C(x) = \log\binom{n}{x}$$

$$D(\theta) = n\log(1-\theta) = n\log q$$

Gamma $\boldsymbol{\gamma}$: b, c, for $\theta = 1/b$ scale parameter

$$A(\theta) = -\theta = -1/b, \quad B(x) = x$$

$$C(x) = (c-1)\log x$$

$$D(\theta) = c\log\theta - \log\Gamma(c) = -c\log b - \log\Gamma(c)$$

Inverse Gaussian $\boldsymbol{I}$: μ, λ, for $\theta = \mu$,

$$A(\theta) = -\frac{\lambda}{2\theta^2} = -\frac{\lambda}{2\mu^2}, \quad B(x) = x$$

$$C(x) = -\tfrac{1}{2}\left[\log\left(\frac{2\pi x^3}{\lambda}\right) + \frac{\lambda}{x}\right]$$

$$D(\theta) = \frac{\lambda}{\theta} = \frac{\lambda}{\mu}$$

Negative binomial $\boldsymbol{NB}$: x, p, for $\theta = p$

$$A(\theta) = \log(1-\theta) = \log(q), \quad B(y) = y$$

$$C(y) = \log\binom{x+y-1}{x-1}$$

$$D(\theta) = x\log(\theta) = x\log(p)$$

Normal $\boldsymbol{N}$: μ, 1, for $\theta = \mu$,

$$A(\theta) = \frac{\theta}{\sigma^2}, \quad B(x) = x$$

$$C(x) = -\frac{1}{2}\left[\log 2(\pi\sigma^2) + \frac{x^2}{\sigma^2}\right]$$

$$D(\theta) = -\frac{\mu^2}{2\sigma^2}$$

Poisson P: λ, for $\theta = \lambda$,

$$\begin{aligned} \Lambda(\theta) &= \log \theta = \log \lambda, \quad B(x) = x \\ C(x) &= -\log(x!) \\ D(\theta) &= -\theta = -\lambda \end{aligned}$$

Families of distributions obtained by sampling from one-parameter exponential families are themselves one-parameter exponential families.

18.3 PARAMETER ESTIMATION

The shared important properties of exponential families enable estimation by likelihood methods, using computer programs such as GenStat (www.vsni.co.uk), R (www.r-project.org), or Stata (www.stata.com).

18.4 GENERALIZED EXPONENTIAL DISTRIBUTIONS

All of the pdfs (or pfs) for the exponential family of distributions satisfy the property

$$\frac{\partial f}{\partial x} = \frac{-g(x)f(x)}{h(x)}, \tag{18.1}$$

where $g(x)$ is a polynomial of degree at most one, and $h(x)$ is a polynomial of degree at most two.

Lye and Martin (1993) propose the univariate generalized exponential family of distributions having densities satisfying Equation (18.1) but where the functions $g(x)$ and $h(x)$ are not constrained to low order polynomial functions.

Generalized Student's *t* Distribution

In particular, Lye and Martin propose the *generalized Student's t distribution*, where

$$g(x) = \sum_{i=0}^{M-1} \alpha_i x^i$$

and

$$h(x) = \gamma^2 x^2.$$

γ^2 is the "degrees of freedom" parameter. When $M = 6$, the pdf has the form

$$f(x) = \exp\left[\theta_1 \tan^{-1}(x/\gamma) + \theta_2 \log(\gamma^2 + x^2) + \sum_{i=3}^{6} \theta_i x^{i-2} - \eta\right],$$

for $-\infty < x < \infty$, where

$$\eta = \int_{-\infty}^{\infty} \exp\left[\theta_1 \tan^{-1}(u/\gamma) + \theta_2 \log(\gamma^2 + u^2) + \sum_{i=3}^{6} \theta_i u^{i-2}\right] du$$

is the normalizing constant.

Variate Relationships

The generalized Student's t variate is denoted by $\boldsymbol{GET} : \gamma, \alpha_0, \alpha_1, \alpha_2, \alpha_3, \alpha_4, \alpha_5$ and satisfies

$$\theta_1 = -\frac{\alpha_0}{\gamma} + \gamma\alpha_2 - \gamma^3\alpha_4$$

$$\theta_2 = -\frac{\alpha_1}{2} + \frac{\gamma^2\alpha_3}{2} - \frac{\gamma^4\alpha_5}{2}$$

$$\theta_3 = -\alpha_2 + \gamma^2\alpha_4$$

$$\theta_4 = \frac{-\alpha_3}{2} + \frac{\gamma^2\alpha_5}{2}$$

$$\theta_5 = -\frac{\alpha_4}{3}$$

$$\theta_6 = -\frac{\alpha_5}{4}.$$

1. The standard Student's t variate $\boldsymbol{t}$: is a special case of the generalized Student's t variate $\boldsymbol{GET}$ with $\nu = \gamma^2$ degrees of freedom where $\theta_2 = -(1 + \gamma^2)$ and $\theta_1 = \theta_3 = \theta_4 = \theta_5 = \theta_6 = 0$.

2. The Cauchy variate $\boldsymbol{C}$: is a special case of the generalized Student's t variate $\boldsymbol{GET}$ with $\theta_2 = -2$ and $\theta_1 = \theta_3 = \theta_4 = \theta_5 = \theta_6 = 0$.

3. The standard Normal variate $\boldsymbol{N} : \mu, \sigma^2$ is a special case of the generalized Student's t variate $\boldsymbol{GET}$ with $\theta_1 = \theta_2 = \theta_5 = \theta_6 = 0$ and $\theta_3 = \mu/\sigma^2$ and $\theta_4 = -1/(1\sigma^2)$.

Generalized Exponential Normal Distribution

The generalized exponential normal variate $\boldsymbol{GEN}$ is defined as a special case of the generalized exponential variate with $\theta_1 = \theta_2 = 0$.

Generalized Lognormal Distribution

In particular, Lye and Martin propose the *generalized lognormal distribution*, where

$$g(x) = \alpha_0 \log(x) + \sum_{i=1}^{M-1} \alpha_i u^{i-1},$$

and

$$h(x) = x.$$

When $M = 6$, the pdf has the form

$$f(x) = \exp\left[\theta_1(\log(x))^2 + \theta_2 \log(x) + \sum_{i=1}^{6} \theta_i x^{i-2}\right],$$

for $x > 0$, where

$$\eta = \int_0^\infty \exp\left[\theta_1(\log(u))^2 + \theta_2 \log(x) + \sum_{i=1}^{6} \theta_i u^{i-2}\right] du$$

is the normalizing constant.

Variate Relationships

$$GEL : \theta_1 = -\alpha_0/2, \quad \theta_2 = -\alpha_1$$
$$\theta_{j+1} = -\alpha_j/(j-1), \quad \text{for } j = 2, 3, 4, 5.$$

1. The standard lognormal variate $\boldsymbol{L} : m, \sigma$ is a special case of the generalized exponential lognormal variate $\boldsymbol{GEL}$ with parameters $\theta_1 = 1/(2\sigma^2), \theta_2 = -\log m/\sigma^2$, and all $theta_j = 0$, for $j = 3, 4, 5$, and 6.

2. The gamma distribution variate $\boldsymbol{\gamma} : b,\ c$ is a special case of the generalized exponential lognormal variate $\boldsymbol{GEL}$ with parameters $\theta_2 = (c-1)$, $\theta_3 = -1/b$, and all $\theta_j = 0$, for $j = 1, 4, 5$, and 6.

3. The Rayleigh distribution variate $\boldsymbol{X} : b$ is a special case of the generalized exponential lognormal variate $\boldsymbol{GEL}$ with parameters $\theta_2 = 1$, $\theta_4 = -1/(2b^2)$, and all $\theta_j = 0$, for $j = 1, 3, 5$, and 6.

Chapter 19

Extreme Value (Gumbel) Distribution

The extreme value distribution was developed as the distribution of the largest of a number of values and was originally applied to the estimation of flood levels. It has since been applied to the estimation of the magnitude of earthquakes. The distribution may also be applied to the study of athletic and other records.

We consider the distribution of the largest extreme. Reversal of the sign of x gives the distribution of the smallest extreme. This is the Type I, the most common of three extreme value distributions, known as the Gumbel distribution.

Variate V: a, b.

Range $-\infty < x < +\infty$.

Location parameter a, the mode.

Scale parameter $b > 0$.

Distribution function	$\exp\{-\exp[-(x-a)/b]\}$
Probability density function	$(1/b)\exp[-(x-a)/b] \times \exp\{-\exp[-(x-a)/b]\}$
Inverse distribution function (of probability α)	$a - b\log[\log(1/\alpha)]$
Inverse survival function (of probability α)	$a - b\log\{\log[1/(1-\alpha)]\}$
Hazard function	$\dfrac{\exp[-(x-a)/b]}{b(\exp\{\exp[-(x-a)/b]\}-1)}$
Moment generating function	$\exp(at)\Gamma(1-bt), \quad t < 1/b$
Characteristic function	$\exp(iat)\Gamma(1-ibt)$

Statistical Distributions, Fourth Edition, by Catherine Forbes, Merran Evans, Nicholas Hastings, and Brian Peacock

Mean	$a - b\Gamma'(1)$ $\Gamma'(1) = -0.57722$ is the first derivative of the gamma function $\Gamma(n)$ with respect to n at $n = 1$.
Variance	$b^2\pi^2/6$
Coefficient of skewness	1.139547
Coefficient of kurtosis	5.4
Mode	a
Median	$a - b\log(\log 2)$

19.1 NOTE

Extreme value variates correspond to the limit, as n tends to infinity, of the maximum value of n-independent random variates with the same continuous distribution. Logarithmic transformations of extreme value variates of Type II (Fréchet) and Type III (Weibull) correspond to Type I Gumbel variates.

Figure 19.1 shows the probability density function for the $\boldsymbol{V} : a, b$ variate, while Figures 19.2 and 19.3 display the corresponding distribution and hazard function, respectively.

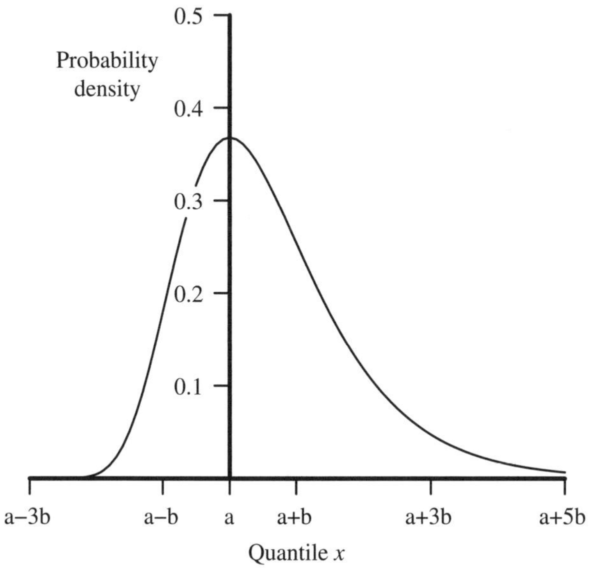

Figure 19.1. Probability density function for the extreme value variate $\boldsymbol{V}$: a, b (largest extreme).

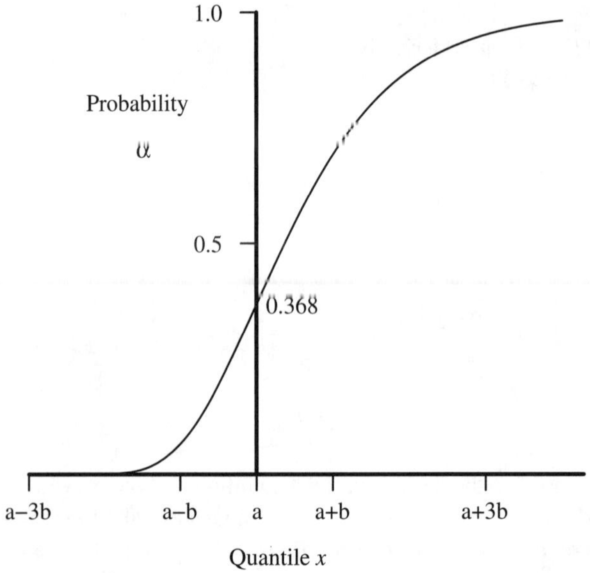

Figure 19.2. Distribution function for the extreme value variate V: a, b (largest extreme).

19.2 VARIATE RELATIONSHIPS

$((V : a, b) - a)/b \sim V : 0, 1$ is the standard Gumbel extreme value variate.

1. The Gumbel extreme value variate V: a, b is related to the exponential variate E: b by

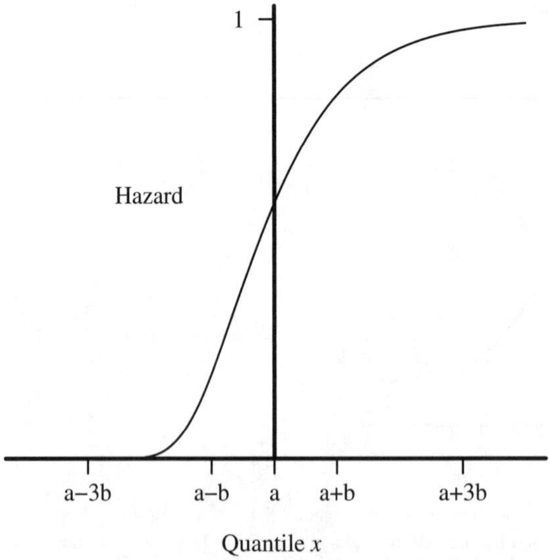

Figure 19.3. Hazard function for the extreme value variate V: a, b (largest extreme).

$$V : a, b = a - \log(\boldsymbol{E} : b).$$

2. Let $(\boldsymbol{E} : b)_i$, $i = 1, \dots, n$, be independent exponential variates with scale parameter b. For large n,

$$(\boldsymbol{E} : b)_{n+a-b\log(m)} \approx \boldsymbol{V} : a, b \quad \text{for } m = 1, 2, \dots$$

3. The standard extreme value variate $\boldsymbol{V}$: 0, 1 is related to the Weibull variate $\boldsymbol{W}$: b, c by

$$-c\log[(\boldsymbol{W} : b, c)/b] \sim \boldsymbol{V} : 0, 1.$$

The extreme value distribution is also known as the "log-Weibull" distribution and is an approximation to the Weibull distribution for large c.

4. The difference of the two independent extreme value variates $(\boldsymbol{V} : a, b)_1$ and $(\boldsymbol{V} : a, b)_2$ is the logistic variate with parameters 0 and b, here denoted $\boldsymbol{X}$: 0, b,

$$\boldsymbol{X} : 0, b \sim (\boldsymbol{V} : a, b)_1 - (\boldsymbol{V} : a, b)_2.$$

5. The standard extreme value variate, $\boldsymbol{V}$: 0, 1 is related to the Pareto variate, here denoted $\boldsymbol{X}$: a, c by

$$\boldsymbol{X} : a, c \sim a\{1 - \exp[-\exp(-\boldsymbol{V} : 0, 1)]\}^{1/c}$$

and the standard power function variate, here denoted $\boldsymbol{X}$: 0, c by

$$\boldsymbol{X} : 0, c \sim \exp\{-\exp[-(\boldsymbol{V} : 0, 1)/c]\}.$$

19.3 PARAMETER ESTIMATION

By the method of maximum likelihood, the estimators $\hat{a}$, $\hat{b}$ are the solutions of the simultaneous equations

$$\hat{b} = \bar{x} - \frac{\sum_{i=1}^{n} x_i \exp\left(\frac{-x_i}{\hat{b}}\right)}{\sum_{i=1}^{n} \exp\left(-\frac{x_i}{\hat{b}}\right)}$$

$$\hat{a} = -\hat{b}\log\left[\frac{1}{n}\sum_{i=1}^{n}\exp\left(\frac{-x_i}{\hat{b}}\right)\right].$$

19.4 RANDOM NUMBER GENERATION

Let $\boldsymbol{R}$ denote a unit rectangular variate. Random numbers of the extreme value variate $\boldsymbol{V}$: a, b can be generated using the relationship

$$\boldsymbol{V} : a, b \sim a - b\log(-\log \boldsymbol{R}).$$

Chapter 20

F (Variance Ratio) or Fisher–Snedecor Distribution

The $\boldsymbol{F}$ variate is the ratio of two chi-squared variates. Chi-squared is the distribution of the variance between data and a theoretical model. The F distribution provides a basis for comparing the ratios of subsets of these variances associated with different factors.

Many experimental scientists make use of the technique called analysis of variance. This method identifies the relative effects of the "main" variables and interactions between these variables. The F distribution represents the ratios of the variances due to these various sources. For example, a biologist may wish to ascertain the relative effects of soil type and water on the yield of a certain crop. The F ratio would be used to compare the variance due to soil type and that due to amount of watering with the residual effects due to other possible causes of variation in yield. The interaction between watering and soil type can also be assessed. The result will indicate which factors, if any, are a significant cause of variation.

Variate $\boldsymbol{F} : \nu, \omega$.

Range $0 \le x < \infty$.

Shape parameters ν, ω, positive integers, referred to as "numerator" and "denominator" degrees of freedom, respectively.

Probability density function $$\frac{\Gamma[(\nu+\omega)/2](\nu/\omega)^{\nu/2}x^{(\nu-2)/2}}{\Gamma(\nu/2)\Gamma(\omega/2)\left[1+(\nu/\omega)x\right]^{(\nu+\omega)/2}}$$

rth Moment about the origin $$\frac{(\omega/\nu)^r\Gamma(\nu/2+r)\Gamma(\omega/2-r)}{\Gamma(\nu/2)\Gamma(\omega/2)}, \quad \omega > 2r$$

Mean $$\frac{\omega}{\omega-2}, \quad \omega > 2$$

Statistical Distributions, Fourth Edition, by Catherine Forbes, Merran Evans, Nicholas Hastings, and Brian Peacock

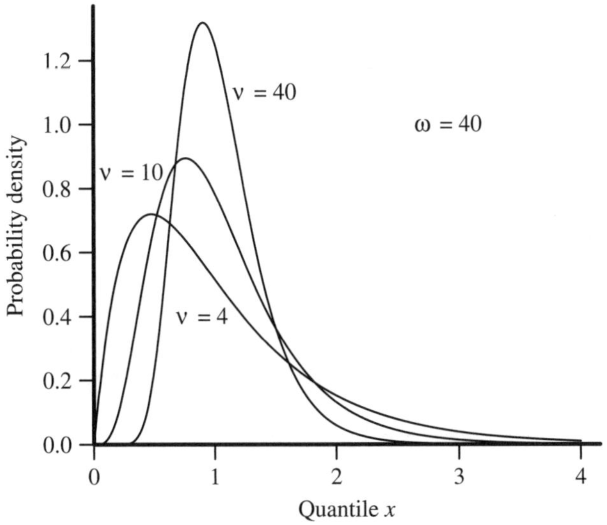

Figure 20.1. Probability density function for the $\boldsymbol{F}$ variate $\boldsymbol{F}: \nu,\ \omega$.

Variance $\dfrac{2\omega^2(\nu+\omega-2)}{\nu(\omega-2)^2(\omega-4)}, \quad \omega > 4$

Mode $\dfrac{\omega(\nu-2)}{\nu(\omega+2)}, \quad \nu > 2$

Coefficient of skewness $\dfrac{(2\nu+\omega-2)[8(\omega-4)]^{1/2}}{\nu^{1/2}(\omega-6)(\nu+\omega-2)^{1/2}}, \quad \omega > 6$

Coefficient of kurtosis $3+\dfrac{12[(\omega-2)^2(\omega-4)+\nu(\nu+\omega-2)(5\omega-22)]}{\nu(\omega-6)(\omega-8)(\nu+\omega-2)}, \quad \omega>8$

Coefficient of variation $\left[\dfrac{2(\nu+\omega-2)}{\nu(\omega-4)}\right]^{1/2}, \quad \omega > 4$

The probability density function of the $\boldsymbol{F}$: ν, 40 variate is shown in Figure 20.1, for selected values of the numerator degrees of freedom parameter, ν. The corresponding distribution functions are shown in Figure 20.2.

20.1 VARIATE RELATIONSHIPS

1. The quantile of the variate $\boldsymbol{F}: \nu, \omega$ at probability level $1-\alpha$ is the reciprocal of the quantile of the variate $\boldsymbol{F}: \omega, \nu$ at probability level α. That is,

$$G_F(1-\alpha : \nu, \omega) = 1/G_F(\alpha : \omega, \nu)$$

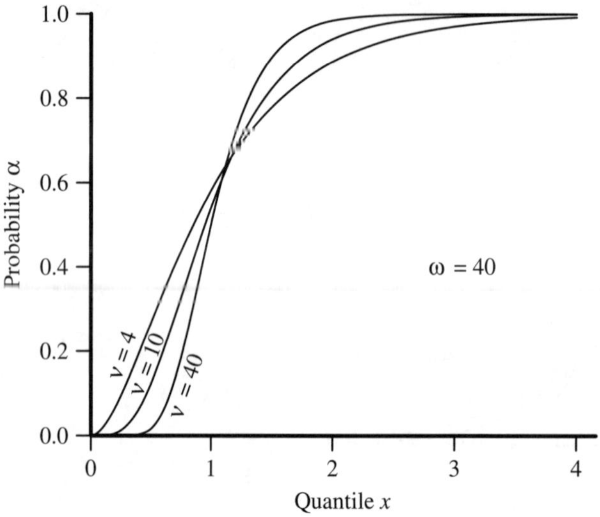

Figure 20.2. Distribution function for the $\boldsymbol{F}$ variate $\boldsymbol{F} : \nu,\ \omega$.

where $G_{\boldsymbol{F}}(\alpha : \nu, \omega)$ is the inverse distribution function of $\boldsymbol{F} : \nu, \omega$ at probability level α.

2. The variate $\boldsymbol{F} : \nu, \omega$ is related to the independent chi-squared variates $\boldsymbol{\chi}^2 : \nu$ and $\boldsymbol{\chi}^2 : \omega$ by

$$\boldsymbol{F} : \nu, \omega \sim \frac{(\boldsymbol{\chi}^2 : \nu)/\nu}{(\boldsymbol{\chi}^2 : \omega)/\omega}.$$

3. As the degrees of freedom ν and ω increase, the $\boldsymbol{F} : \nu, \omega$ variate tends to normality.

4. The variate $\boldsymbol{F} : \nu, \omega$ tends to the chi-squared variate $\boldsymbol{\chi}^2 : \nu$ as ω tends to infinity:

$$\boldsymbol{F} : \nu, \omega \approx (1/\nu)(\boldsymbol{\chi}^2 : \nu) \quad \text{as} \quad \omega \to \infty.$$

5. The quantile of the variate $\boldsymbol{F}$: $1, \omega$ at probability level α is equal to the square of the quantile of the Student's $\boldsymbol{t}$ variate $\boldsymbol{t} : \omega$ at probability level $\frac{1}{2}(1+\alpha)$. That is,

$$G_{\boldsymbol{F}}(\alpha : 1, \omega) = \left[G_t \left(\frac{1}{2}(1+\alpha) : \omega \right) \right]^2.$$

where G is the inverse distribution function. In terms of the inverse survival function the relationship is

$$Z_{\boldsymbol{F}}(\alpha : 1, \omega) = \left[Z_t \left(\frac{1}{2}\alpha : \omega \right) \right]^2.$$

6. The variate $\boldsymbol{F} : \nu, \omega$ and the beta variate $\boldsymbol{\beta} : w/2, \nu/2$ are related by

$$\begin{aligned}\Pr[(\boldsymbol{F} : \nu, \omega) > x] &= \Pr[(\boldsymbol{\beta} : \omega/2, \nu/2) \leq \omega/(\omega + \nu x)] \\ &= S_{\boldsymbol{F}}(x : \nu, \omega) \\ &= F_{\boldsymbol{\beta}}([\omega/(\omega + \nu x)] : \omega/2, \nu/2)\end{aligned}$$

where S is the survival function and F is the distribution function. Hence the inverse survival function $Z_{\boldsymbol{F}}(\alpha : \nu, \omega)$ of the variate $\boldsymbol{F} : \nu, \omega$ and the inverse distribution function $G_{\boldsymbol{\beta}}(\alpha : \omega/2, \nu/2)$ of the beta variate $\boldsymbol{\beta} : \omega/2, \nu/2$ are related by

$$\begin{aligned}Z_{\boldsymbol{F}}(\alpha : \nu, \omega) &= G_{\boldsymbol{F}}((1 - \alpha) : \nu, \omega) \\ &= (\omega/\nu)\{[1/G_{\boldsymbol{\beta}}(\alpha : \omega/2, \nu/2)] - 1\}\end{aligned}$$

where α denotes probability.

7. The variate $\boldsymbol{F} : \nu, \omega$ and the inverted beta variate $\boldsymbol{I\beta} : \nu/2, \omega/2$ are related by

$$\boldsymbol{F} : \nu, \omega \sim (\omega/\nu)(\boldsymbol{I\beta} : \nu/2, \omega/2).$$

8. Consider two sets of independent normal variates $(\boldsymbol{N} : \mu_1, \sigma_1)_i$; $i = 1, \dots, n_1$ and $(\boldsymbol{N} : \mu_2, \sigma_2)_j$; $j = 1, \dots, n_2$. Define variates $\bar{x}_1,\ \bar{x}_2,\ s_1^2,\ s_2^2$ as follows:

$$\bar{x}_1 = \sum_{i=1}^{n_1} (\boldsymbol{N} : \mu_1, \sigma_1)_i/n_1, \quad \bar{x}_2 = \sum_{j=1}^{n_2} (\boldsymbol{N} : \mu_2, \sigma_2)_j/n_2$$

$$s_1^2 = \sum_{i=m}^{n_1} [(\boldsymbol{N} : \mu_1, \sigma_1)_i - \bar{x}_1]^2/n_1$$

$$s_2^2 = \sum_{j=1}^{n_2} [(\boldsymbol{N} : \mu_2, \sigma_2)_j - \bar{x}_2]^2/n_2.$$

Then

$$\boldsymbol{F} : n_1, n_2 \sim \frac{n_1 s_1^2/(n_1 - 1)\sigma_1^2}{n_2 s_2^2/(n_2 - 1)\sigma_2^2}.$$

9. The variate $\boldsymbol{F} : \nu, \omega$ is related to the binomial variate with Bernoulli trial parameter $\frac{1}{2}(\omega + \nu - 2)$ and Bernoulli probability parameter p by

$$\Pr\left[(\boldsymbol{F} : \nu, \omega) < \frac{\omega p}{\nu(1 - p)}\right] = 1 - \Pr\left[\left(\boldsymbol{B} : \frac{1}{2}(\omega + \nu - 2), p\right) \leq \frac{1}{2}(\nu - 2)\right]$$

where $\omega + \nu$ is an even integer.

10. The ratio of two independent Laplace variates, with parameters 0 and b, denoted $(\boldsymbol{L}:0,b)_i$, $i=1,2$ is related to the $\boldsymbol{F}$: 2, 2 variate by

$$\boldsymbol{F}:2,2 \sim \frac{|(\boldsymbol{L}:0,b)_1|}{|(\boldsymbol{L}:0,b)_2|}.$$

Chapter 21

F (Noncentral) Distribution

Variate F: ν, ω, δ.

Range $0 < x < \infty$.

Shape parameters ν, ω, positive integers are the degrees of freedom, and $\delta > 0$ the noncentrality parameter.

Probability density function
$$k\frac{\exp(-\delta/2)\nu^{\nu/2}\omega^{\omega/2}x^{(\nu-2)/2}}{\text{B}(\nu/2,\omega/2)(\omega+\nu x)^{(\nu+\omega)/2}},$$
$$\text{where } k = 1 + \sum_{j=1}^{\infty}\left(\frac{(\nu\delta x)/2}{\omega+\nu x}\right)^j$$
$$\times\frac{(\nu+\omega)(\nu+\omega+2)(\nu+\omega+2j-2)}{j!\nu(\nu+2)\cdots(\nu+2j-2)}$$

rth Moment about the origin
$$\left(\frac{\omega}{\nu}\right)^r\frac{\Gamma((\nu/2)+r)\Gamma((\omega/2)-r)}{\Gamma(\omega/2)}\sum_{j=0}^{r}\frac{\binom{r}{j}\left(\frac{\delta\nu}{2}\right)^j}{\Gamma\left(\frac{\nu}{2}+j\right)}$$

Mean
$$\frac{\omega(\nu+\delta)}{\nu(\omega-2)},\quad \omega>2$$

Variance
$$2\left(\frac{\omega}{\nu}\right)^2\left(\frac{(\nu+\delta)^2+(\nu+2\delta)(\omega-2)}{(\omega-2)^2(\omega-4)}\right),\quad \omega>4$$

Mean deviation
$$\frac{2[(\nu+\delta)^2+(\nu+2\delta)(\omega-2)]}{[(\nu+\delta)^2(\omega-4)]^{1/2}},\quad \omega>2$$

The probability density function of the $\boldsymbol{F}$: 4, 40, δ variate for selected values of the noncentrality parameter δ is shown in Figure 21.1.

Statistical Distributions, Fourth Edition, by Catherine Forbes, Merran Evans, Nicholas Hastings, and Brian Peacock

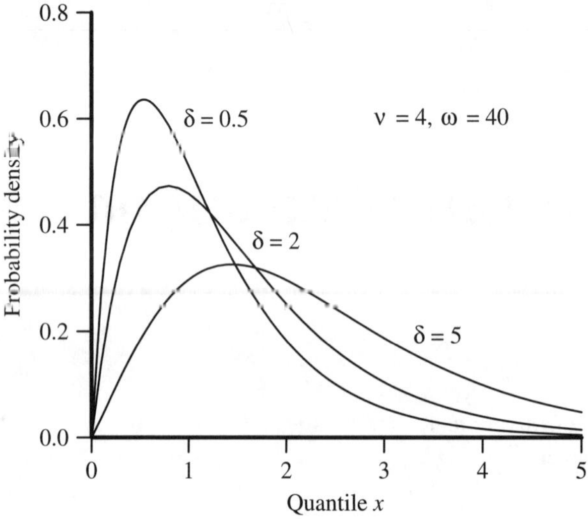

Figure 21.1. Probability density function for the (noncentral) F variate F: ν, ω, δ.

21.1 VARIATE RELATIONSHIPS

1. The noncentral F variate F: ν, ω, δ is related to the independent noncentral chi-squared variate $\chi^2 : \nu, \delta$ and central chi-squared variate $\chi^2 : \omega$ by

$$F : \nu, \omega, \delta \sim \frac{(\chi^2 : \nu, \delta)/\nu}{(\chi^2 : \omega)/\omega}.$$

2. The noncentral F variate $F : \nu, \omega, \delta$ tends to the (central) F variate $F : \nu, \omega$ as δ tends to zero.

3. If the negative binomial variate $NB : \omega/2, p$, and the Poisson variate $P : \delta/2$ are independent, then they are related to the noncentral F variate $F : \nu, \omega, \delta$ (for ν even) by

$$\Pr[(F : \nu, \omega, \delta) < p\omega/\nu]$$
$$= \Pr[[(NB : \omega/2, p) - (P : \delta/2)] \geq \nu/2].$$

Chapter 22

Gamma Distribution

The gamma distribution includes the chi-squared, Erlang, and exponential distributions as special cases, but the shape parameter of the gamma is not confined to integer values. The gamma distribution starts at the origin and has a flexible shape. The parameters are easy to estimate by matching moments.

Variate $\boldsymbol{\gamma}$: b, c.

Range $0 \leq x < \infty$.

Scale parameter $b > 0$. Alternative parameter λ, $\lambda = 1/b$.

Shape parameter $c > 0$.

Distribution function	For c an integer see Erlang distribution.
Probability density function	$(x/b)^{c-1}[\exp(-x/b)]/b\Gamma(c)$, where $\Gamma(c)$ is the gamma function with argument c (see Section 8.1)
Moment generating function	$(1 - bt)^{-c}, \quad t < 1/b$
Laplace transform of the pdf	$(1 + bs)^{-c}, \quad s > -1/b$
Characteristic function	$(1 - ibt)^{-c}$
Cumulant generating function	$-c \log(1 - ibt)$
rth Cumulant	$(r - 1)!cb^r$
rth Moment about the origin	$b^r\Gamma(c + r)/\Gamma(c)$
Mean	bc
Variance	b^2c
Mode	$b(c - 1), \quad c \geq 1$
Coefficient of skewness	$2c^{-1/2}$
Coefficient of kurtosis	$3 + 6/c$
Coefficient of variation	$c^{-1/2}$

Statistical Distributions, Fourth Edition, by Catherine Forbes, Merran Evans, Nicholas Hastings, and Brian Peacock

The probability density function of the γ: 1, c variate, for selected values of the shape parameter c, is shown in Figure 22.1, with the corresponding distribution functions shown in Figure 22.2. The hazard functions for the γ: 1, 0.5 and γ: 1, 2 variates are shown in Figure 22.3.

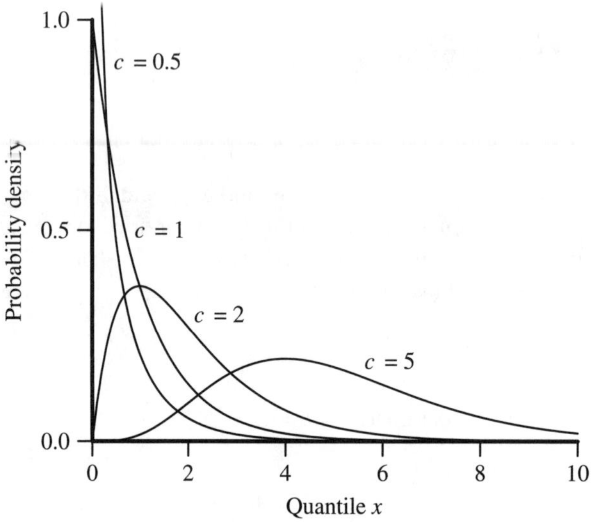

Figure 22.1. Probability density function for the gamma variate $\gamma : 1,\ c$.

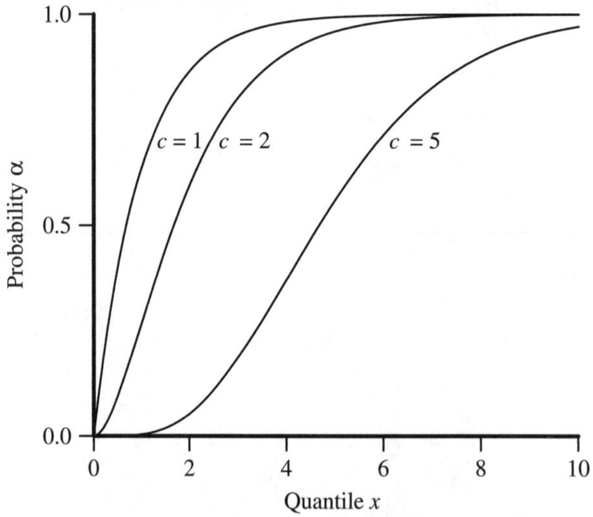

Figure 22.2. Distribution function for the gamma variate γ: 1, c.

22.1 VARIATE RELATIONSHIPS

$(\gamma : b, c)/b \sim \gamma : 1, c$, standard gamma variate.

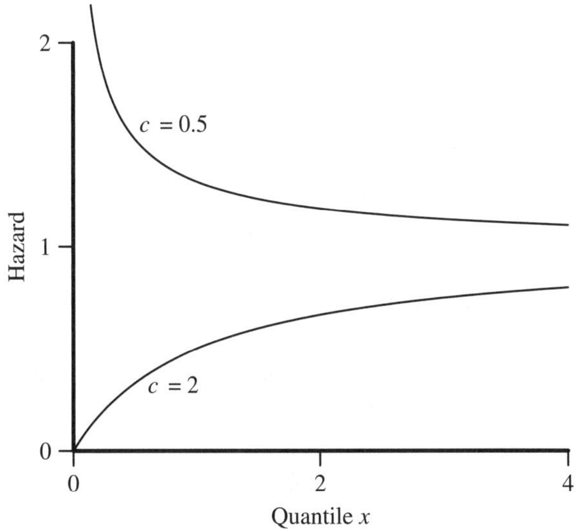

Figure 22.3. Hazard function for the gamma variate $\boldsymbol{\gamma}$: 1, c.

1. If $\boldsymbol{E}$: b is an exponential variate with mean b, then

$$\boldsymbol{\gamma} : b, 1 \sim \boldsymbol{E} : b.$$

2. If the shape parameter c is an integer, the gamma variate $\boldsymbol{\gamma} : 1, c$ is also referred to as the Erlang variate.

3. If the shape parameter c is such that $2c$ is an integer, then

$$\boldsymbol{\gamma} : 1, c \sim \frac{1}{2}(\boldsymbol{\chi}^2 : 2c)$$

where $\boldsymbol{\chi}^2 : 2c$ is a chi-squared variate with $2c$ degrees of freedom.

4. The sum of n-independent gamma variates with shape parameters c_i is a gamma variate with shape parameter $c = \sum_{i=1}^{n} c_i$

$$\sum_{i=1}^{n} (\boldsymbol{\gamma} : b, c_i) \sim \boldsymbol{\gamma} : b, c, \quad \text{where} \quad c = \sum_{i=1}^{n} c_i.$$

5. The independent standard gamma variates with shape parameters c_1 and c_2 are related to the beta variate with shape parameters c_1, c_2, denoted $\boldsymbol{\beta} : c_1, c_2$, by

$$(\boldsymbol{\gamma} : 1, c_1)/[(\boldsymbol{\gamma} : 1, c_1) + (\boldsymbol{\gamma} : 1, c_2)] \sim \boldsymbol{\beta} : c_1, c_2.$$

22.2 PARAMETER ESTIMATION

Parameter	Estimator	Method
Scale parameter, b	$s^2/\bar{x}$	Matching moments
Shape parameter, c	$(\bar{x}/s)^2$	Matching moments

Maximum-likelihood estimators $\hat{b}$ and $\hat{c}$ are solutions of the simultaneous equations [see Section 8.1 for $\psi(c)$].

$$\hat{b} = \bar{x}/\hat{c}$$

$$\log \hat{c} - \psi(\hat{c}) = \log \left[\bar{x} \bigg/ \left(\prod_{i=1}^{n} x_i \right)^{1/n} \right].$$

22.3 RANDOM NUMBER GENERATION

Variates $\boldsymbol{\gamma}$: b, c for the case where c is an integer (equivalent to the Erlang variate) can be computed using

$$\boldsymbol{\gamma} : b, c \sim -b \log \left(\prod_{i=1}^{c} \boldsymbol{R}_i \right) = \sum_{i=1}^{c} -b \log \boldsymbol{R}_i.$$

where the $\boldsymbol{R}_i$ are independent unit rectangular variates.

22.4 INVERTED GAMMA DISTRIBUTION

The variate $1/(\boldsymbol{\gamma} : \lambda = 1/b, c)$ is the inverted gamma variate and has probability density function (with quantile y)

$$\frac{\exp(-\lambda/y)\lambda^c (1/y)^{c+1}}{\Gamma(c)}.$$

Its mean is $\lambda/(c-1)$ for $c > 1$ and its variance is

$$\lambda^2/[(c-1)^2(c-2)] \quad \text{for } c > 2.$$

22.5 NORMAL GAMMA DISTRIBUTION

For a normal $N : \mu, \sigma$ variate, the normal gamma prior density for (μ, σ) is obtained by specifying a normal density for the conditional prior of μ given σ, and an inverted gamma density for the marginal prior of σ, and is

$$\frac{\tau^{1/2}}{(2\pi)^{1/2}\sigma} \exp\left(-\frac{\tau}{2\sigma^2}(\mu - \mu_0)^2 \right)$$
$$\times \frac{2}{\Gamma(\nu/2)} \left(\frac{\nu s^2}{2} \right)^{\omega/2} \frac{1}{\sigma^{\nu+1}} \exp\left(-\frac{\nu s^2}{2\sigma^2} \right)$$

where μ_0, τ, ν, and s^2 are the parameters of the prior distribution. In particular,

$$E(\mu|\sigma) = E(\mu) = \mu_0, \qquad \text{variance} \quad (\mu|\sigma) = \sigma^2/\tau.$$

This is often used as a tractable conjugate prior distribution in Bayesian analysis. (See Section 6.5.)

22.6 GENERALIZED GAMMA DISTRIBUTION

Variate $\boldsymbol{\gamma} : a, b, c, k$.

Range $0 < a < x$.

Location parameter $a > 0$. Scale parameter $b > 0$.

Shape parameters $c > 0$ and $k > 0$.

Probability density function	$\dfrac{k(x-a)^{kc-1}}{b^{kc}\Gamma(c)} \exp\left[-\left(\dfrac{x-a}{b}\right)^k\right]$
rth Moment about a	$b^r\Gamma(c + r/k)/\Gamma(c)$
Mean	$a + b\Gamma(c + 1/k)/\Gamma(c)$
Variance	$b^2\{\Gamma(c + 2/k)/\Gamma(c) - [\Gamma(c + 1/k)/\Gamma(c)]^2\}$
Mode	$a + b(c - 1/k)^{1/k}, \quad c > 1/k$

Variate Relationships

1. Special cases of the generalized gamma variate $\boldsymbol{\gamma}$: a, b, c, k are the following:

Gamma variate $\boldsymbol{\gamma} : b, c$ with $k = 1, a = 0$.

Exponential variate $\boldsymbol{E}$: b with $c = k = 1, a = 0$.

Weibull variate $\boldsymbol{W}$: b, k with $c = 1, a = 0$.

Chi-squared variate $\boldsymbol{\chi}^2 : \nu$ with $a = 0, b = 2, c = \nu/2, k = 1$.

2. The generalized and standard gamma variates are related by

$$\left(\frac{(\boldsymbol{\gamma} : a, b, c, k) - a}{b}\right)^{1/k} \sim \boldsymbol{\gamma} : 1, c.$$

3. The generalized gamma variate $\boldsymbol{\gamma} : a, b, c, k$ tends to the lognormal variate $\boldsymbol{L}$: m, σ when k tends to zero, c tends to infinity, and b tends to infinity such that k^2c tends to $1/\sigma^2$ and $bc^{1/k}$ tends to m.

4. The generalized gamma variate $\boldsymbol{\gamma} : 0, b, c, k$ with $a = 0$ tends to the power function variate with parameters b and p when c tends to zero and k tends to infinity such that ck tends to p, and tends to the Pareto variate with parameters b and p when c tends to zero and k tends to minus infinity such that ck tends to $-p$.

Chapter 23

Geometric Distribution

Suppose we were interviewing a series of people for a job and we had established a set of criteria that must be met for a candidate to be considered acceptable. The geometric distribution would be used to describe the number of interviews that would have to be conducted in order to get the first acceptable candidate.

Variate $\boldsymbol{G}: p$.

Quantile n, number of trials.

Range $n \geq 0$, n an integer.

Parameter p, the Bernoulli probability parameter, $0 < p < 1$.

Given a sequence of independent Bernoulli trials, where the probability of success at each trial is p, the geometric variate $\boldsymbol{G}: p$ is the number of trials or failures before the first success. Let $q = 1 - p$.

Distribution function	$1 - q^{n+1}$
Probability function	pq^n
Inverse distribution function (of probability α)	$[\log(1-\alpha)/\log(q)] - 1$, rounded down, for $p \leq \alpha \leq 1$
Inverse survival function (of probability α)	$[\log(\alpha)/\log(q)] - 1$, rounded down, for $0 \leq \alpha \leq q$
Moment generating function	$p/[1 - q\exp(t)]$, $t < -\log(q)$
Probability generating function	$p/(1 - qt)$, $\quad \lvert t \rvert < 1/q$
Characteristic function	$p/[1 - q\exp(it)]$
Mean	q/p

Statistical Distributions, Fourth Edition, by Catherine Forbes, Merran Evans, Nicholas Hastings, and Brian Peacock

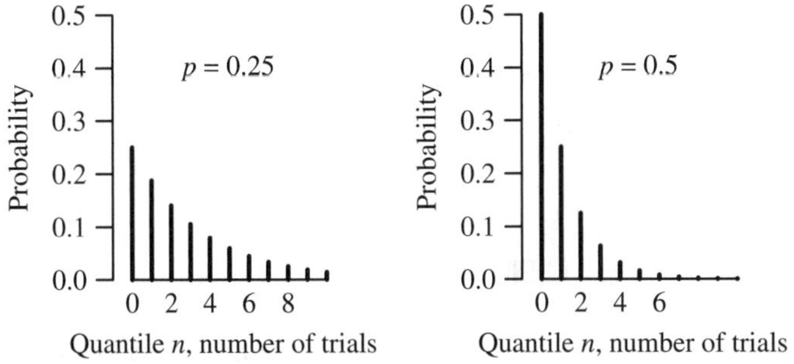

Figure 23.1. Probability function for the geometric variate G: p.

Moments about mean	
Variance	q/p^2
Third	$q(1+q)/p^3$
Fourth	$(9q^2/p^4) + (q/p^2)$
Mode	0
Coefficient of skewness	$(1+q)/q^{1/2}$
Coefficient of kurtosis	$9 + p^2/q$
Coefficient of variation	$q^{-1/2}$

23.1 NOTES

1. The geometric distribution is a discrete analogue of the continuous exponential distribution and only these are characterized by a "lack of memory."

2. An alternative form of the geometric distribution involves the number of trials up to and including the first success. This has probability function pq^{n-1}, mean $1/p$, and probability generating function $pt/(1-qt)$. The geometric distribution is also sometimes called the Pascal distribution.

The probability functions for the $\boldsymbol{G}$: 0.25 and $\boldsymbol{G}$: 0.5 variates are shown in the two panels of Figure 23.1.

23.2 VARIATE RELATIONSHIPS

1. The geometric variate is a special case of the negative binomial variate $\boldsymbol{NB}$: x, p with $x = 1$.

$$\boldsymbol{G} : p \sim \boldsymbol{NB} : 1, p.$$

2. The sum of x-independent geometric variates is the negative binomial variate

$$\sum_{i=1}^{x} (\boldsymbol{G} : p)_i \sim \boldsymbol{NB} : x, p.$$

23.3 RANDOM NUMBER GENERATION

Random numbers of the geometric variate $\boldsymbol{G}$: p can be generated from random numbers of the unit rectangular variate $\boldsymbol{R}$ using the relationship

$$\boldsymbol{G} : \mathrm{p} \sim [\log(\boldsymbol{R})/\log(1 - p)] - 1$$

rounded up to the next larger integer.

Chapter 24

Hypergeometric Distribution

Suppose a wildlife biologist is interested in the reproductive success of wolves that had been introduced into an area. Her approach could be to catch a sample, size X, and place radio collars on them. The next year, after the wolf packs had been allowed to spread, a second sample, size n, could be caught and the number of this sample that had the radio collars would be x. The hypergeometric distribution could then be used to estimate the total number of wolves, N, in the area. This example illustrates an important point in the application of theory to practice—that is, the assumptions that must be made to make the application of a particular theory (distribution) reliable and valid. In the cited example it was assumed that the wolves had intermixed randomly and that the samples were drawn randomly and independently on the successive years. Also, it was assumed that there had been minimal losses due to the activities of hunters or farmers or gains due to reproduction or encroachment from other areas. Probability and statistical distribution theory provide useful research tools, which must be complemented by domain knowledge.

Variate $\boldsymbol{H}$: $N,\ X,\ n$.

Quantile x, number of successes.

Range $\max[0, n - N + X] \le x \le \min[X, n]$.

Parameters N, the number of elements in the population; X, the number of successes in the population; n, sample size.

From a population of N elements of which X are successes (i.e., possess a certain attribute) we draw a sample of n items without replacement. The number of successes in such a sample is a hypergeometric variate $\boldsymbol{H}$: $N,\ X,\ n$.

Probability function (probability of exactly x successes)	$\binom{X}{x}\binom{N-X}{n-x}\Big/\binom{N}{n}$
Mean	nX/N
Moments about the mean	
Variance	$(nX/N)(1 - X/N)(N - n)/(N - 1)$

Statistical Distributions, Fourth Edition, by Catherine Forbes, Merran Evans, Nicholas Hastings, and Brian Peacock

Third $\dfrac{(nX/N)(1 - X/N)(1 - 2X/N)(N - n)(N - 2n)}{(N - 1)(N - 2)}$

Fourth $\dfrac{(nX/N)(1 - X/N)(N - n)}{(N - 1)(N - 2)(N - 3)} \times \left\{N(N + 1) - 6n(N - n)\right.$

$\left. +(3X/N)(1 - X/N)\left[n(N - n)(N + 6) - 2N^2\right]\right\}$

Coefficient of skewness $\dfrac{(N - 2X)(N - 1)^{1/2}(N - 2n)}{[nX(N - X)(N - n)]^{1/2}(N - 2)}$

Coefficient of kurtosis $\left[\dfrac{N^2(N - 1)}{n(N - 2)(N - 3)(N - n)}\right]$

$\times \left[\dfrac{N(N + 1) - 6n(N - n)}{X(N - X)}\right.$

$\left. + \dfrac{3n(N - n)(N + 6)}{N^2} - 6\right]$

Coefficient of variation $\{(N - X)(N - n)/nX(N - 1)\}^{1/2}$

24.1 NOTE

Successive values of the probability function $f(x)$ are related by

$$f(x + 1) = f(x)(n - x)(X - x)/[(x + 1)(N - n - X + x + 1)]$$
$$f(0) = (N - X)!(N - n)!/[(N - X - n)!N!].$$

24.2 VARIATE RELATIONSHIPS

1. The hypergeometric variate $\boldsymbol{H}$: $N,\ X,\ n$ can be approximated by the binomial variate with Bernoulli probability parameter $p = X/N$ and Bernoulli trial parameter n, denoted $\boldsymbol{B}$: $n,\ p$, provided $n/N < 0.1$, and N is large. That is, when the sample size is relatively small, the effect of nonreplacement is slight.

2. The hypergeometric variate $\boldsymbol{H}$: $N,\ X,\ n$ tends to the Poisson variate $\boldsymbol{P}$: λ as X, N, and n all tend to infinity for X/N small and nX/N tending to λ. For large n, but X/N not too small, it tends to a normal variate.

24.3 PARAMETER ESTIMATION

Parameter	Estimation	Method/Properties
N	max integer $\leq nX/x$	Maximum likelihood
X	max integer $\leq (N + 1)x/n$	Maximum likelihood
X	Nx/n	Minimum variance, unbiased

24.4 RANDOM NUMBER GENERATION

To generate random numbers of the hypergeometric variate $\boldsymbol{H}$: $N,\ X,\ n$, select n-independent, unit rectangular random numbers $\boldsymbol{R}_i, i = 1, \ldots, n$. If $\boldsymbol{R}_i < p_i$ record a success, where

$$p_1 = X/N$$

$$p_{i+1} = [(N - i + 1)p_i - d_i]/(N - i), \quad i \geq 2$$

where

$$d_i = \begin{cases} 0 & \text{if } \boldsymbol{R}_i \geq p_i \\ 1 & \text{if } \boldsymbol{R}_i < p_i \end{cases}.$$

24.5 NEGATIVE HYPERGEOMETRIC DISTRIBUTION

If two items of the corresponding type are replaced at each selection (see Section 7.3), the number of successes in a sample of n items is the negative hypergeometric variate with parameters $N,\ X,\ n$. The probability function is

$$\binom{X + x - 1}{x}\binom{N - X + n - x - 1}{n - x}\Bigg/\binom{N + n - 1}{n}$$

$$= \binom{-X}{x}\binom{-N + X}{n - x}\Bigg/\binom{-N}{n}.$$

The mean is nX/N and the variance is $(nX/N)(1 - X/N) \times (N + n)/(N + 1)$. This variate corresponds to the beta binomial beta variate with integer parameters $\nu = X, \omega = N - X$.

The negative hypergeometric variate with parameters $N,\ X,\ n$ tends to the binomial variate, $\boldsymbol{B}$: $n,\ p$, as N and X tend to infinity and X/N to p, and to the negative binomial variate, $\boldsymbol{NB}$: $n,\ p$, as N and n tend to infinity and $N/(N + n)$ to p.

24.6 GENERALIZED HYPERGEOMETRIC DISTRIBUTION

A generalization, with parameters $N,\ X,\ n$ taking any real values, forms an extensive class, which includes many well-known discrete distributions and which has attractive features. (See Johnson, Kemp and Kotz, 2005, p. 259.)

Chapter 25

Inverse Gaussian (Wald) Distribution

The inverse Gaussian distribution has applications in the study of diffusion processes and as a lifetime distribution model.

Variate $\boldsymbol{I}$: μ, λ.

Range $x > 0$.

Location parameter $\mu > 0$, the mean.

Scale parameter $\lambda > 0$.

Probability density function	$\left(\lambda/(2\pi x^3)\right)^{1/2} \exp\left(\dfrac{-\lambda(x-\mu)^2}{2\mu^2 x}\right)$
Moment generating function	$\exp\left\{\dfrac{\lambda}{\mu}\left[1-\left(1-\dfrac{2\mu^2 t}{\lambda}\right)^{1/2}\right]\right\}$
Characteristic function	$\exp\left\{\dfrac{\lambda}{\mu}\left[1-\left(1-\dfrac{2\mu^2 it}{\lambda}\right)^{1/2}\right]\right\}$
rth Cumulant	$1 \cdot 3 \cdot 5 \cdots (2r-3)\mu^{2r-1}\lambda^{1-r},\ r \geq 2$
Cumulant generating function	$\dfrac{\lambda}{\mu}\left[1-\left(1+2\dfrac{\mu^2 it}{\lambda}\right)^{1/2}\right]$
rth Moment about the origin	$\mu^r \sum_{i=0}^{r-1} \dfrac{(r-1+i)!}{i!(r-1-i)!}\left(\dfrac{\mu}{2\lambda}\right)^i,\ r \geq 2$
Mean	μ
Variance	μ^3/λ

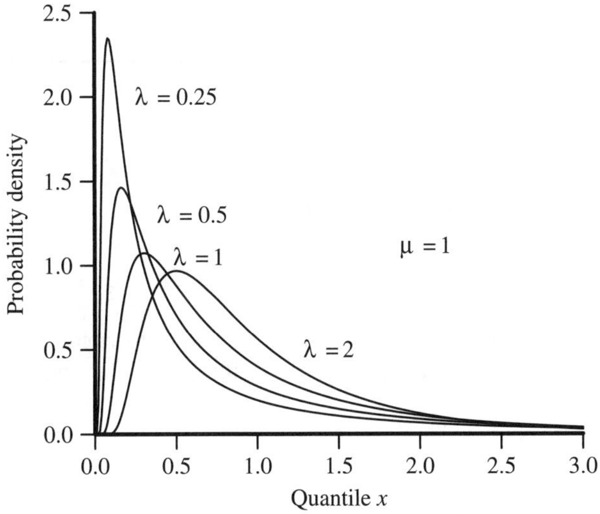

Figure 25.1. Probability density function for the inverse Gaussian variate I: μ, λ.

Mode	$\mu\left[\left(1+\frac{9\mu^2}{4\lambda^2}\right)^{1/2}-\frac{3\mu}{2\lambda}\right]$
Coefficient of skewness	$3(\mu/\lambda)^{1/2}$
Coefficient of kurtosis	$3+15\mu/\lambda$
Coefficient of variation	$(\mu/\lambda)^{1/2}$

The probability density function for the I: 1, λ variate is shown in Figure 25.1 for selected values of the scale parameter λ.

25.1 VARIATE RELATIONSHIPS

1. The standard Wald variate is a special case of the inverse Gaussian variate I: μ, λ, for $\mu = 1$.
2. The standard inverse Gaussian variate I: μ, λ tends to the standard normal variate N: 0, 1 as λ tends to infinity.

25.2 PARAMETER ESTIMATION

Parameter	Estimator	Method/Properties
μ	$\bar{x}$	Maximum likelihood
λ	$n\bigg/\left(\sum_{i=1}^{n} x_i^{-1} - (\bar{x})^{-1}\right)$	Maximum likelihood
λ	$(n-1)\bigg/\left(\sum_{i=1}^{n} x_i^{-1} - (\bar{x})^{-1}\right)$	Minimum variance,unbiased

Chapter 26

Laplace Distribution

The Laplace distribution is often known as the double-exponential distribution. In modeling, the Laplace provides a heavier tailed alternative to the normal distribution.

Variate $\boldsymbol{L}$: $a,\ b$.

Range $-\infty < \boldsymbol{x} < \infty$.

Location parameter $-\infty < \boldsymbol{a} < \infty$, the mean.

Scale parameter $b > 0$.

Distribution function	$\frac{1}{2}\exp\left[-\left(\frac{a-x}{b}\right)\right], \quad x < a$
	$1-\frac{1}{2}\exp\left[-\left(\frac{x-a}{b}\right)\right], \quad x \geq a$
Probability density function	$\frac{1}{2b}\exp\left(-\frac{\lvert x-a\rvert}{b}\right)$
Moment generating function	$\frac{\exp(at)}{1-b^2t^2}, \quad \lvert t\rvert < b^{-1}$
Characteristic function	$\frac{\exp(iat)}{1+b^2t^2}$
rth Cumulant	$\begin{cases} 2(r-1)!b^r, & r \text{ even} \\ 0, & r \text{ odd} \end{cases}$
Mean	a
Median	a
Mode	a
rth Moment about the mean, μ_r	$\begin{cases} r!b^r, & r \text{ even} \\ 0, & r \text{ odd} \end{cases}$
Variance	$2b^2$

Statistical Distributions, Fourth Edition, by Catherine Forbes, Merran Evans, Nicholas Hastings, and Brian Peacock

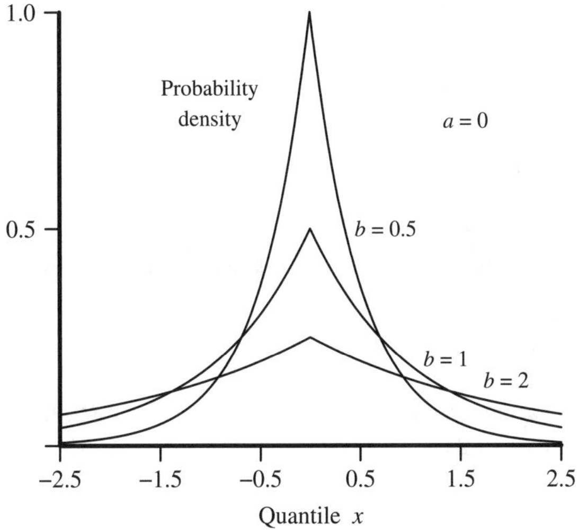

Figure 26.1. Probability density function for the Laplace variate $\boldsymbol{L} : a, b$.

Coefficient of skewness	0
Coefficient of kurtosis	6
Coefficient of variation	$2^{1/2}\left(\frac{b}{a}\right)$

The probability density function for the $\boldsymbol{L} : 0, b$ variate is shown in Figure 26.1 for selected values of the scale parameter b. The corresponding probability functions are shown in Figure 26.2.

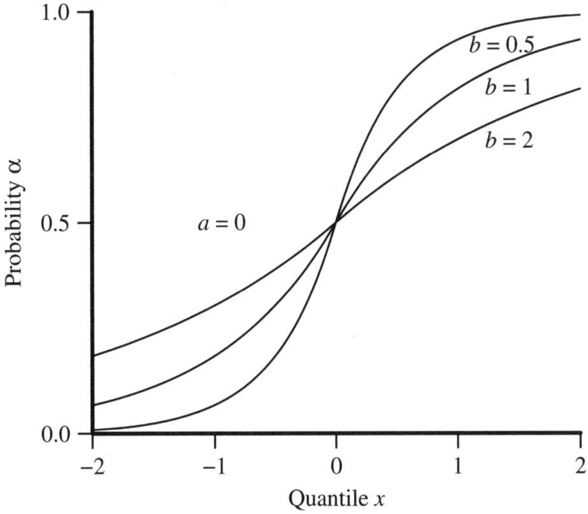

Figure 26.2. Distribution function for the Laplace variate $\boldsymbol{L} : a, b$.

26.1 VARIATE RELATIONSHIPS

1. The Laplace variate **L**: a, b is related to the independent exponential variates **E**: b and **E**: 1 by

$$\boldsymbol{E} : b \sim |(\boldsymbol{L} : a, b) - a|$$

$$\boldsymbol{E} : 1 \sim |(\boldsymbol{L} : a, b) - a|/b.$$

2. The Laplace variate **L**: 0, b is related to two independent exponential variates **E**: b by

$$\boldsymbol{L} : 0, b \sim (\boldsymbol{E} : b)_1 - (\boldsymbol{E} : b)_2.$$

3. Two independent Laplace variates, with parameter $a = 0$, are related to the **F** variate with parameters $\nu = \omega = 2$, **F**: 2, 2, by

$$\boldsymbol{F} : 2, 2 \sim |(\boldsymbol{L} : 0, b)_1/(\boldsymbol{L} : 0, 2)_2|.$$

26.2 PARAMETER ESTIMATION

Parameter	Estimator	Method/Properties
a	Median	Maximum likelihood
b	$\frac{1}{n}\sum_{i=1}^{n}\lvert x_i - a\rvert$	Maximum likelihood

26.3 RANDOM NUMBER GENERATION

The standard Laplace variate **L**: 0, 1 is related to the independent unit rectangular variates $\boldsymbol{R}_1$, $\boldsymbol{R}_2$ by

$$\boldsymbol{L} : 0, 1 \sim \log(\boldsymbol{R}_1/\boldsymbol{R}_2).$$

Chapter 27

Logarithmic Series Distribution

Range $x \geq 1$, an integer.

Shape parameter $0 < c < 1$.

For simplicity, also let $k = -1/\log(1-c)$.

Probability function	kc^x/x
Probability generating function	$\log(1-ct)/\log(1-c), \quad \lvert t \rvert < 1/c$
Moment generating function	$\log[1-c\exp(t)]/\log(1-c)$
Characteristic function	$\log[1-c\exp(it)]/\log(1-c)$
Moments about the origin	
Mean	$kc(1-c)$
Second	$kc/(1-c)^2$
Third	$kc(1+c)/(1-c)^3$
Fourth	$kc(1+4c+c^2)/(1-c)^4$
Moments about the mean	
Variance	$kc(1-kc)/(1-c)^2$
Third	$kc(1+c-3kc+2k^2c^2)/(1-c)^3$
Fourth	$\dfrac{kc[1+4c+c^2-4kc(1+c)+6k^2c^2-3k^3c^3]}{(1-c)^4}$
Coefficient of skewness	$\dfrac{(1+c)-3kc+2k^2c^2}{(kc)^{1/2}(1-kc)^{3/2}}$
Coefficient of kurtosis	$\dfrac{1+4c+c^2-4kc(1+c)+6k^2c^2-3k^3c^3}{kc(1-kc)^2}$

Statistical Distributions, Fourth Edition, by Catherine Forbes, Merran Evans, Nicholas Hastings, and Brian Peacock

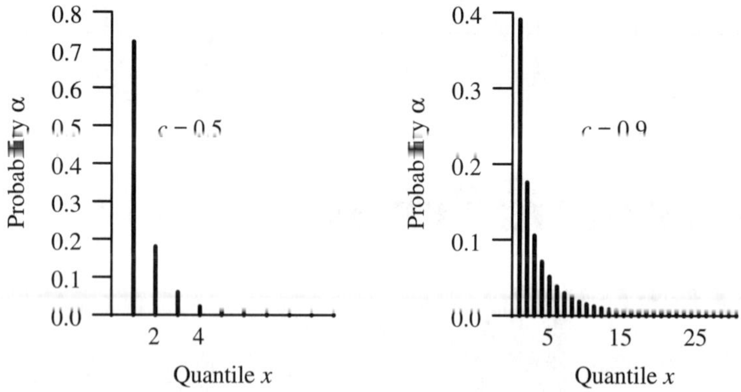

Figure 27.1. Probability function for the logarithmic series variate.

The probability functions for the logarithmic series variate, with shape parameters $c = 0.5$ and $c = 0.9$ are shown in the two panels of Figure 27.1.

27.1 VARIATE RELATIONSHIPS

1. The logarithmic series variate with parameter c corresponds to the power series distribution variate with parameter c and series function $-\log(1-c)$.
2. The limit toward zero of a zero truncated (i.e., excluding $x = 0$) negative binomial variate with parameters x and $p = 1 - c$ is a logarithmic series variate with parameter c.

27.2 PARAMETER ESTIMATION

The maximum-likelihood and method of moments estimators $\hat{c}$ satisfy the equation

$$\bar{x} = \frac{\hat{c}}{-(1-\hat{c})\log(1-\hat{c})}.$$

Other asymptotically unbiased estimators of c are

$$1 - \begin{pmatrix}\text{proportion of} \\ \text{x's equal to 1}\end{pmatrix} \bigg/ \bar{x}$$

$$1 - \left(n^{-1}\sum x_j^2\right) \Big/ \bar{x}.$$

Chapter 28

Logistic Distribution

The distribution function of the logistic is used as a model for growth. For example, with a new product we often find that growth is initially slow, then gains momentum, and finally slows down when the market is saturated or some form of equilibrium is reached.

Applications include the following:

- Market penetration of a new product
- Population growth
- The expansion of agricultural production
- Weight gain in animals

Range $-\infty < x < \infty$.

Location parameter a, the mean.

Scale parameter $b > 0$.

Alternative parameter $k = \pi b/3^{1/2}$, the standard deviation.

Distribution function

$$1 - \{1 + \exp[(x - a)/b]\}^{-1}$$
$$= \{1 + \exp[-(x - a)/b]\}^{-1}$$
$$= \frac{1}{2}\left\{1 + \tanh\left[\frac{1}{2}(x - a)/b\right]\right\}$$

Probability density function

$$\frac{\exp[-(x - a)/b]}{b\{1 + \exp[-(x - a)/b]\}^2}$$
$$= \frac{\exp[(x - a)/b]}{b\{1 + \exp[(x - a)/b]\}^2}$$
$$= \frac{\operatorname{sech}^2[(x - a)/2b]}{4b}$$

Statistical Distributions, Fourth Edition, by Catherine Forbes, Merran Evans, Nicholas Hastings, and Brian Peacock

Inverse distribution function (of probability α)	$a + b\log[\alpha/(1-\alpha)]$
Survival function	$\{1 + \exp[(x-a)/b]\}^{-1}$
Inverse survival function (of probability α)	$a + b\log[(1-\alpha)/\alpha]$
Hazard function	$\{b\{1 + \exp[-(x-a)/b]\}\}^{-1}$
Cumulative hazard function	$\log\{1 + \exp[(x-a)/b]\}$
Moment generating function	$\exp(at)\Gamma(1-bt)\Gamma(1+bt) = \pi bt\exp(at)/\sin(\pi bt)$
Characteristic function	$\exp(iat)\pi bit/\sin(\pi bit)$
Mean	a
Variance	$\pi^2 b^2/3$
Mode	a
Median	a
Coefficient of skewness	0
Coefficient of kurtosis	4.2
Coefficient of variation	$\pi b/(3^{1/2}a)$

28.1 NOTES

1. The logistic distribution is the limiting distribution, as n tends to infinity, of the average of the largest to smallest sample values, of random samples of size n from an exponential-type distribution.
2. The standard logistic variate, here denoted X: 0, 1 with parameters $a = 0$, $b = 1$, has a distribution function F_X and probability density function f_X with the properties

$$f_X = F_X(1 - F_X)$$
$$x = \log[F_X/(1 - F_X)].$$

The probability density function for the logistic variate is shown in Figure 28.1 for location parameter $a = 0$ and for selected values of the scale parameter b. The corresponding probability functions are shown in Figure 28.2.

28.2 VARIATE RELATIONSHIPS

The standard logistic variate, here denoted X: 0, 1, is related to the logistic variate, denoted X: a, b by

$$X : 0, 1 \sim [(X : a, b) - a]/b.$$

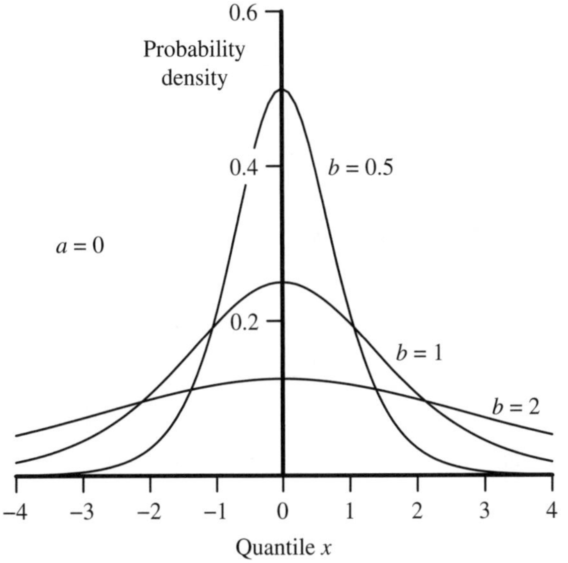

Figure 28.1. Probability density function for the logistic variate.

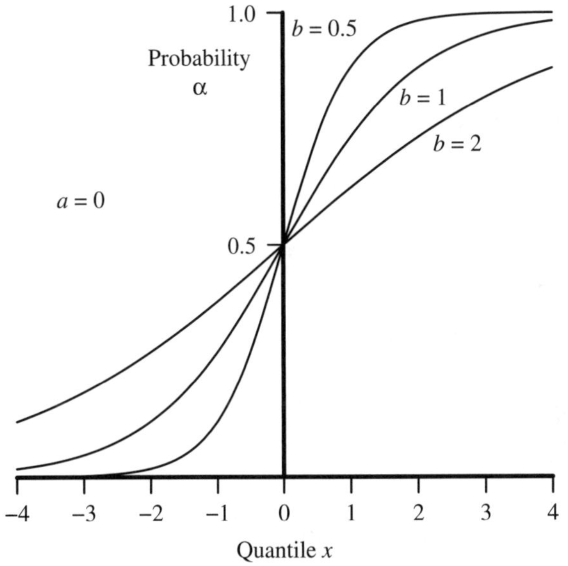

Figure 28.2. Distribution function for the logistic variate.

1. The standard logistic variate $\boldsymbol{X}$: 0, 1 is related to the standard exponential variate $\boldsymbol{E}$: 1 by

$$\boldsymbol{X} : 0, 1 \sim -\log\{\exp(-\boldsymbol{E} : 1)/[1 + \exp(-\boldsymbol{E} : 1)]\}.$$

For two independent standard exponential variates $\boldsymbol{E}$: 1, then

$$X : 0, 1 \sim -\log[(\boldsymbol{E} : 1)_1/(\boldsymbol{E} : 1)_2].$$

2. The standard logistic variate $\boldsymbol{X}$: 0, 1 is the limiting form of the weighted sum of n independent standard Gumbel extreme value variates $\boldsymbol{V}$: 0, 1 as n tends to infinity

$$X : 0, 1 \approx \sum_{i=1}^{n} (V : 0, 1)_i / i, \quad n \to \infty.$$

3. Two independent standard Gumbel extreme value variates, $\boldsymbol{V}$: *a, b*, are related to the logistic variate $\boldsymbol{X}$: 0, *b* by

$$X : 0, b \sim (V : a, b)_1 - (V : a, b)_2.$$

4. The Pareto variate, here denoted $\boldsymbol{Y}$: *a, c*, is related to the standard logistic variate $\boldsymbol{X}$: 0, 1 by

$$X : 0, 1 \sim -\log\{[(Y : a, c)/a]^c - 1\}.$$

5. The standard power function variate, here denoted $\boldsymbol{Y}$: 1, *c*, is related to the standard logistic variate $\boldsymbol{X}$: 0, 1 by

$$X : 0, 1 \sim -\log[(Y : 1, c)^{-c} - 1].$$

28.3 PARAMETER ESTIMATION

The maximum-likelihood estimators $\hat{a}$ and $\hat{b}$ of the location and scale parameters are the solutions of the simultaneous equations

$$\frac{n}{2} = \sum_{i=1}^{n} \frac{\exp\left(\frac{x_i - \hat{a}}{\hat{b}}\right)}{\left[1 + \exp\left(\frac{x_i - \hat{a}}{\hat{b}}\right)\right]}$$

$$n = \sum_{i=1}^{n} \left(\frac{x_i - \hat{a}}{\hat{b}}\right) \frac{\exp\left(\frac{x_i - \hat{a}}{\hat{b}}\right) - 1}{\exp\left(\frac{x_i - \hat{a}}{\hat{b}}\right) + 1}.$$

28.4 RANDOM NUMBER GENERATION

Let $\boldsymbol{R}$ denote a unit rectangular variate. Random numbers of the logistic variate $\boldsymbol{X}$: *a*, *b* can be generated using the relation

$$X : a, b \sim a + b \log[R/(1 - R)].$$

Chapter 29

Lognormal Distribution

The lognormal distribution is applicable to random variables that are constrained by zero but have a few very large values. The resulting distribution is asymmetrical and positively skewed. Examples include the following:

- The weight of adults
- The concentration of minerals in deposits
- Duration of time off due to sickness
- Distribution of wealth
- Machine down times

The application of a logarithmic transformation to the data can allow the data to be approximated by the symmetrical normal distribution, although the absence of negative values may limit the validity of this procedure.

Variate $\boldsymbol{L} : m, \sigma$ or $\boldsymbol{L} : \mu, \sigma$.

Range $0 \leq x < \infty$.

Scale parameter $m > 0$, the median.

Alternative parameter μ, the mean of log L.

m and μ are related by $m = \exp \mu$, $\mu = \log m$.

Shape parameter $\sigma > 0$, the standard deviation of log L.

For compactness the substitution $\omega = \exp(\sigma^2)$ is used in several formulas.

Probability density function $$\frac{1}{x\sigma(2\pi)^{1/2}} \exp\left(\frac{-[\log(x/m)]^2}{2\sigma^2}\right)$$

$$= \frac{1}{x\sigma(2\pi)^{1/2}} \exp\left(\frac{-(\log x - \mu)^2}{2\sigma^2}\right)$$

Statistical Distributions, Fourth Edition, by Catherine Forbes, Merran Evans, Nicholas Hastings, and Brian Peacock

rth Moment about the origin	$m^r \exp\left(\frac{1}{2}r^2\sigma^2\right)$
	$= \exp\left(r\mu + \frac{1}{2}r^2\sigma^2\right)$
Mean	$m \exp\left(\frac{1}{2}\sigma^2\right)$
Variance	$m^2\omega(\omega - 1)$
Mode	m/ω
Median	m
Coefficient of skewness	$(\omega + 2)(\omega - 1)^{1/2}$
Coefficient of kurtosis	$\omega^4 + 2\omega^3 + 3\omega^2 - 3$
Coefficient of variation	$(\omega - 1)^{1/2}$

The probability density function of the $\boldsymbol{L}: 0, \sigma$ variate, for selected values of the shape parameter σ, is shown in Figure 29.1, with the corresponding distribution and hazard functions shown in Figures 29.2 and 29.3, respectively.

29.1 VARIATE RELATIONSHIPS

1. The lognormal variate with median m and with σ denoting the standard deviation of $\log \boldsymbol{L}$ is expressed by $\boldsymbol{L}$: m, σ. (Alternatively, if μ, the mean of $\log \boldsymbol{L}$, is used as a parameter, the lognormal variate is expressed by $\boldsymbol{L}$: μ, σ.) The lognormal variate is related to the normal variate with mean μ and standard deviation σ, denoted $\boldsymbol{N}$:

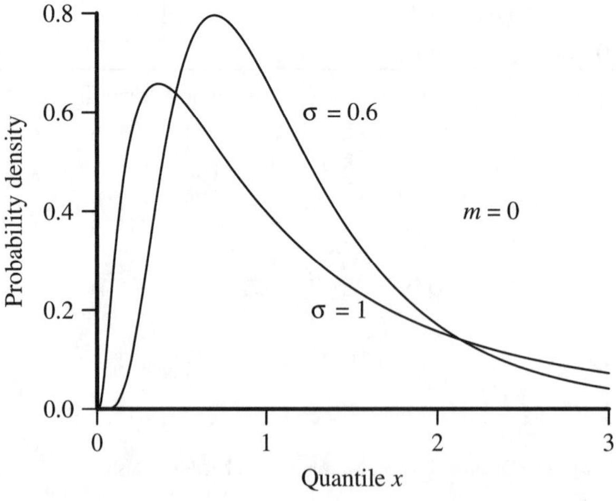

Figure 29.1. Probability density function for the lognormal variate $\boldsymbol{L}$: m, σ.

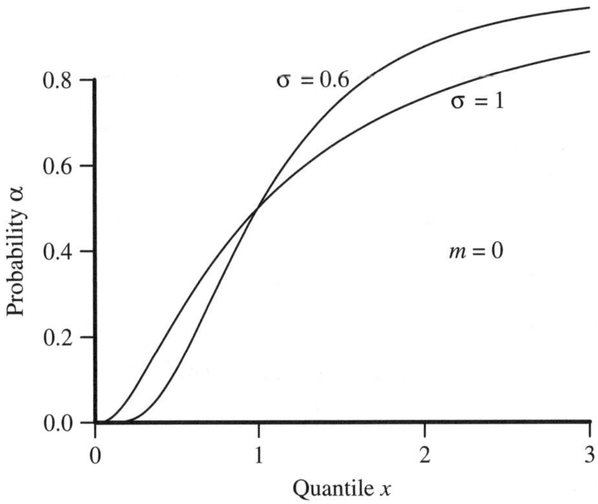

Figure 29.2. Distribution function for the lognormal variate $\boldsymbol{L}: m,\ \sigma$.

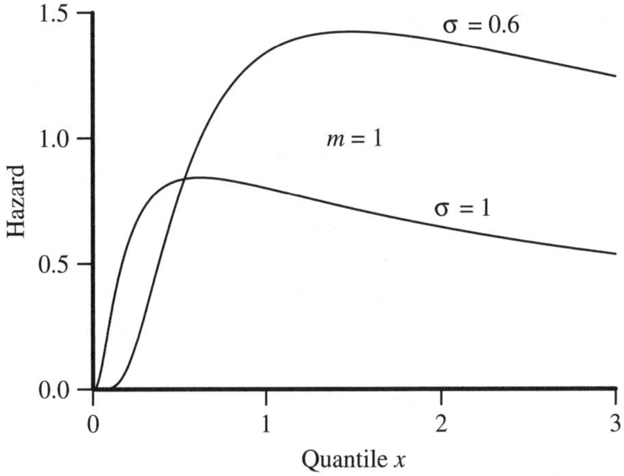

Figure 29.3. Hazard function for the lognormal variate $\boldsymbol{L}: m,\ \sigma$.

μ, σ, by the following:

$$\begin{aligned}
\boldsymbol{L} : m, \sigma &\sim \exp(\boldsymbol{N} : \mu, \sigma) \sim \exp[\mu + \sigma(\boldsymbol{N} : 0, 1)] \\
&\sim m \exp(\sigma \boldsymbol{N} : 0, 1) \\
\log(\boldsymbol{L} : m, \sigma) &\sim (\boldsymbol{N} : \mu, \sigma) \sim \mu + \sigma(\boldsymbol{N} : 0, 1) \\
\Pr[(\boldsymbol{L} : \mu, \sigma) \leq x] &= \Pr[(\exp(\boldsymbol{N} : \mu, \sigma)) \leq x] \\
&= \Pr[(\boldsymbol{N} : \mu, \sigma) \leq \log x] \\
&= \Pr[(\boldsymbol{N} : 0, 1) \leq \log((x - \mu)/\sigma)]
\end{aligned}$$

2. For small σ, the normal variate $N : \log \mu, \sigma$ approximates the lognormal variate $L : \mu, \sigma$.

3. Transformations of the following form, for a and b constant, of the lognormal variate $L : \mu, \sigma$ are also lognormal:

$$\exp(a)(L : \mu, \sigma)^b \sim L : a + b\mu, b\sigma.$$

4. For two independent lognormal variates, $L : \mu_1, \sigma_1$ and $L : \mu_2, \sigma_2$,

$$(L : \mu_1, \sigma_1) \times (L : \mu_2, \sigma_2) \sim L : \mu_1 + \mu_2, (\sigma_1^2 + \sigma_2^2)^{1/2}$$
$$(L : \mu_1, \sigma_1)/(L : \mu_2, \sigma_2) \sim L : \mu_1 - \mu_2, (\sigma_1^2 + \sigma_2^2)^{1/2}.$$

5. The geometric mean of the n-independent lognormal variates $L : \mu, \sigma$ is also a lognormal variate

$$\left(\prod_{i=1}^{n} (L : \mu, \sigma)_i\right)^{1/n} \sim L : \mu, \sigma/n^{1/2}.$$

29.2 PARAMETER ESTIMATION

The following estimators are derived by transformation to the normal distribution.

Parameter	Estimator
Median, m	$\hat{m} = \exp \hat{\mu}$
Mean of log(L), μ	$\hat{\mu} = \left(\frac{1}{n}\right) \sum_{i=1}^{n} \log x_i$
Variance of log(L), σ^2	$\hat{\sigma}^2 = \left(\frac{1}{n-1}\right) \sum_{i=1}^{n} [\log(x_i - \hat{\mu})]^2$

29.3 RANDOM NUMBER GENERATION

The relationship of the lognormal variate $L : m, \sigma$ to the unit normal variate N: 0, 1 gives

$$L : m, \sigma \sim m \exp(\sigma N : 0, 1)$$
$$\sim \exp[\mu + \sigma(N : 0, 1)].$$

Chapter 30

Multinomial Distribution

The multinomial variate is a multidimensional generalization of the binomial. Consider a trial that can result in only one of k possible distinct outcomes, labeled $A_i, i = 1, \ldots, k$. Outcome A_i occurs with probability p_i. The multinomial distribution relates to a set of n-independent trials of this type. The multinomial multivariate is $\boldsymbol{M} = [\boldsymbol{M}_i]$, where $\boldsymbol{M}_i$ is the variate "number of times event A_i occurs," $i = 1, \ldots, k$. The quantile is a vector $\boldsymbol{x} = [x_1, \ldots, x_k]'$. For the multinomial variate, x_i is the quantile of $\boldsymbol{M}_i$ and is the number of times event A_i occurs in the n trials.

Suppose we wish to test the robustness of a complex component of an automobile under crash conditions. The component may be damaged in various ways each with different probabilities. If we wish to evaluate the probability of a particular combination of failures we could apply the multinomial distribution. A useful approximation to the multinomial distribution is the application of the chi-squared distribution to the analysis of contingency tables.

Multivariate $\boldsymbol{M} : n, p_1, \ldots, p_k$.

Range $x_i \geq 0,\ \sum_{i=1}^k x_i = n$, x_i an integer.

Parameters n and $p_i, i = 1, \ldots, k$, where $0 < p_i < 1,\ \sum_{i=1}^k p_i = 1$.

The joint probability function $f(x_1, \ldots, x_k)$ is the probability that each event A_i occurs x_i times, $i = 1, \ldots, k$, in the n trials, and is given by

Probability function	$n! \prod_{i=1}^{k} (p_i^{x_i}/x_i!)$
Probability generating function	$\left(\sum_{i=1}^{k} p_i t_i\right)^n$
Moment generating function	$\left(\sum_{i=1}^{k} p_i \exp(t_i)\right)^n$
Cumulant generating function	$n \log \left(\sum_{i=1}^{k} p_i \exp(it_i)\right)$

Statistical Distributions, Fourth Edition, by Catherine Forbes, Merran Evans, Nicholas Hastings, and Brian Peacock

Individual elements, $\boldsymbol{M}_i$

Mean	np_i	
Variance	$np_i(1-p_i)$	
Covariance	np_ip_j	$i \neq j$

$$\text{Third cumulant} \quad \begin{cases} np_i(1-p_i)(1-2p_i) & i=j=k \\ -np_ip_k(1-2p_i) & i=j\neq k \\ 2np_ip_jp_k & i,\ j,\ k \text{ all distinct} \end{cases}$$

$$\text{Fourth cumulant} \quad \begin{cases} np_i(1-p_i)\left[1-6p_i(1-p_i)\right] & i=j=k=l \\ -np_ip_l\left[1-6p_i(1-p_i)\right] & i=j=k\neq l \\ -np_ip_k[1-2p_i-2p_k+6p_ip_k] & i=j\neq k=l \\ 2np_ip_kp_l(1-3p_i) & i=j\neq k\neq l \\ -6np_ip_jp_kp_l & i,\ j,\ k,\ l \text{ all distinct} \end{cases}$$

30.1 VARIATE RELATIONSHIPS

If $k=2$ and $p_1=p$, the multinomial variate corresponds to the binomial variate $\boldsymbol{B}$: n, p. The marginal distribution of each $\boldsymbol{M}_i$ is the binomial distribution with parameters n, p_i.

30.2 PARAMETER ESTIMATION

For individual elements

Parameter	Estimator	Method/Properties
p_i	x_i/n	Maximum likelihood

Chapter 31

Multivariate Normal (Multinormal) Distribution

A multivariate normal distribution is a multivariate extension of the normal distribution.

The bivariate normal distribution may be applied to the description of correlated variables, such as smoking and performance on respiratory function tests. An extension of the concept may be applied to multiple correlated variables such as heart disease, body weight, exercise, and dietary habits.

Multivariate $\boldsymbol{MN}$: $\boldsymbol{\mu}$, $\boldsymbol{\Sigma}$.

Quantile $\boldsymbol{x} = [x_1, \ldots, x_k]'$ a $k \times 1$ vector.

Range $-\infty < x_i < \infty$, for $i = 1, \ldots, k$.

Location parameter, the $k \times 1$ mean vector, $\boldsymbol{\mu} = [\mu_1, \ldots, \mu_k]'$, with $-\infty < \mu_i < \infty$.

Parameter $\boldsymbol{\Sigma}$, the $k \times k$ positive definite variance-covariance matrix, with elements $\Sigma_{ij} = \sigma_{ij}$. Note $\sigma_{ii} = \sigma_i^2$.

Probability density function	$f(\boldsymbol{x}) = (2\pi)^{-(1/2)k}\lvert\boldsymbol{\Sigma}\rvert^{-1/2} \times \exp\left[-\frac{1}{2}(\boldsymbol{x}-\boldsymbol{\mu})'\boldsymbol{\Sigma}^{-1}(\boldsymbol{x}-\boldsymbol{\mu})\right]$
Characteristic function	$\exp\left(-\frac{1}{2}\boldsymbol{t}'\boldsymbol{\Sigma}\boldsymbol{t}\right)\exp\left(it'\boldsymbol{\mu}\right)$
Moment generating function	$\exp\left(\boldsymbol{\mu}'t + \frac{1}{2}t'\boldsymbol{\Sigma}t\right)$
Cumulant generating function	$-\frac{1}{2}t'\boldsymbol{\Sigma}t + i't'\boldsymbol{\mu}$
Mean	$\boldsymbol{\mu}$
Variance-covariance	$\boldsymbol{\Sigma}$
Moments about the mean	
Third	0
Fourth	$\sigma_{ij}\sigma_{kl} + \sigma_{ik}\sigma_{jl} + \sigma_{il}\sigma_{jk}$

Statistical Distributions, Fourth Edition, by Catherine Forbes, Merran Evans, Nicholas Hastings, and Brian Peacock

rth Cumulant	0 for $r > 2$
For individual elements $\boldsymbol{MN}_i$	
Probability density function	$(2\pi)^{-1/2}\lvert\Sigma_{ii}\rvert^{-1/2} \times \exp\left[-\frac{1}{2}(x_i-\mu_i)'\Sigma_{ii}^{-1}(x_i-\mu_i)\right]$
Mean	μ_i
Variance	$\Sigma_{ii} = \sigma_i^2$
Covariance	$\Sigma_{ij} = \sigma_{ij}$

31.1 VARIATE RELATIONSHIPS

1. A fixed linear transformation of a multivariate normal variate is also a multivariate normal variate. For $\boldsymbol{a}$, a constant $j\times 1$ vector, and $\boldsymbol{B}$, a $j \times k$ fixed matrix, the resulting variate is of dimension $j\times 1$:

$$\boldsymbol{a} + \boldsymbol{B}(\boldsymbol{MN} : \boldsymbol{\mu}, \boldsymbol{\Sigma}) \sim (\boldsymbol{MN} : \boldsymbol{a} + \boldsymbol{B\mu}, \boldsymbol{B\Sigma B}').$$

2. The multinormal variate with $k = 1$ corresponds to the normal variate $\boldsymbol{N} : \mu, \sigma$, where $\mu = \mu_1$ and $\sigma^2 = \Sigma_{11}$.

3. The sample mean of variates with any joint distribution with finite mean and variance tends to the multivariate normal form. This is the simplest form of the multivariate central limit theorem.

31.2 PARAMETER ESTIMATION

For individual elements

Parameter	Estimator	Method/Properties
μ_i	$\bar{x}_i = \sum_{t=1}^{n} x_{ti}$	Maximum likelihood
Σ_{ij}	$\sum_{t=1}^{n}(x_{ti} - \bar{x}_i)(x_{tj} - \bar{x}_j)$	Maximum likelihood

Chapter 32

Negative Binomial Distribution

The Pascal variate is the number of failures before the xth success in a sequence of Bernoulli trials, where the probability of success at each trial is p and the probability of failure is $q = 1 - p$. This generalizes to the negative binomial variate for noninteger x.

Suppose prosecution and defense lawyers were choosing 12 citizens to comprise a jury. The Pascal distribution could be applied to estimate the number of rejections before the jury selection process was completed. The Pascal distribution is an extension of the geometric distribution, which applies to the number of failures before the first success. The Pascal distribution generalizes to the negative binomial, when the definition of "success" is not an integer. An example of the negative binomial is the number of scoops of ice cream needed to fill a bowl, as this is not necessarily an integer.

Variate ***NB***: $x,\ p$.

Quantile y.

Range $0 \leq y < \infty$, y an integer.

Parameters $0 < x < \infty, 0 < p < 1, q = 1 - p$.

Distribution function (Pascal)	$\sum_{i=0}^{y} \binom{x+i-1}{x-1} p^x q^i$ (integer x only)
Probability function (Pascal)	$\binom{x+y-1}{x-1} p^x q^y$ (integer x only)
Probability function	$\Gamma(x+y)p^x q^y / \Gamma(x) y!$
Moment generating function	$p^x(1 - q\exp t)^{-x}, t < \log(q)$
Probability generating function	$p^x(1 - qt)^{-x}, \lvert t \rvert < 1/q$
Characteristic function	$p^x\left[1 - q\exp(it)\right]^{-x}$
Cumulant generating function	$x\log(p) - x\log\left[1 - q\exp(it)\right]$

Statistical Distributions, Fourth Edition, by Catherine Forbes, Merran Evans, Nicholas Hastings, and Brian Peacock

Cumulants	
First	xq/p
Second	xq/p^2
Third	$xq(1+q)/p^3$
Fourth	$xq(6q+p^2)/p^4$
Mean	xq/p
Moments about the mean	
Variance	xq/p^2
Third	$xq(1+q)/p^3$
Fourth	$(xq/p^4)(3xq+6q+p^2)$
Coefficient of skewness	$(1+q)(xq)^{-1/2}$
Coefficient of kurtosis	$3+6/x+p^2/(xq)$
Coefficient of variation	$(xq)^{-1/2}$
Factorial moment generating function	$(1-qt/p)^{-x}$
rth Factorial moment about the origin	$(q/p)^r\Gamma(x+r)/\Gamma(x)$

32.1 NOTE

The Pascal variate is a special case of the negative binomial variate with integer values only. An alternative form of the Pascal variate involves trials up to and including the xth success.

The probability function for the ***NB*** : x, p variate is shown in Figures 32.1, 32.2 and 32.3 for selected values of x and p.

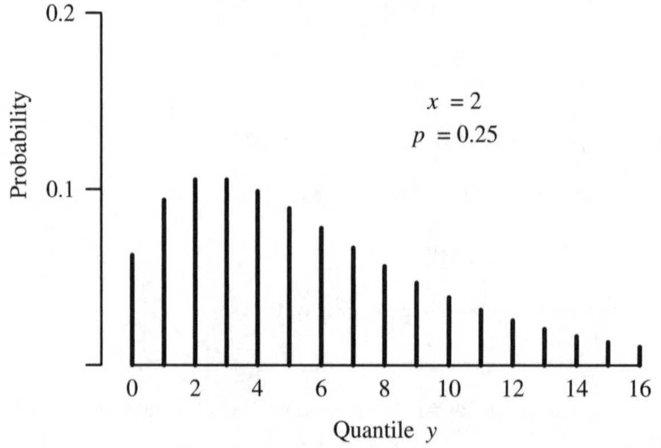

Figure 32.1. Probability function for the negative binomial variate ***NB***: x, p.

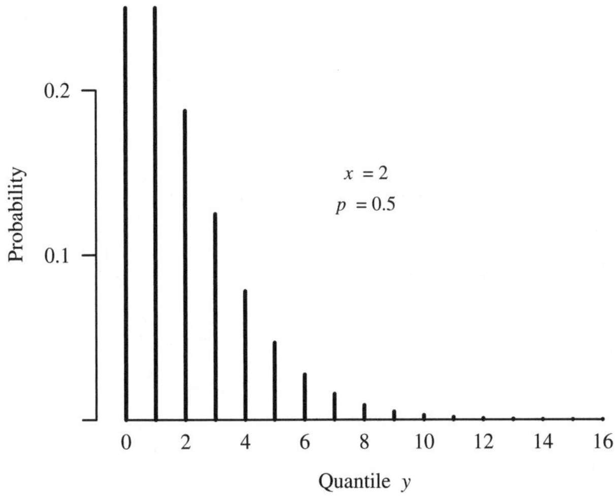

Figure 32.2. Probability function for the negative binomial variate ***NB***: *x, p*.

32.2 VARIATE RELATIONSHIPS

1. The sum of k-independent negative binomial variates $\boldsymbol{NB}$: x_i, $p; i = 1, \ldots, k$ is a negative binomial variate $\boldsymbol{NB}$: x', p, where

$$\sum_{i=1}^{k}(\boldsymbol{NB} : x_i, p) \sim \boldsymbol{NB} : x', p, \quad \text{where} \quad x' = \sum_{i=1}^{k} x_i.$$

2. The geometric variate $\boldsymbol{G}$: p is a special case of the negative binomial variate with $x = 1$.

$$\boldsymbol{G} : p \sim \boldsymbol{NB} : 1, p$$

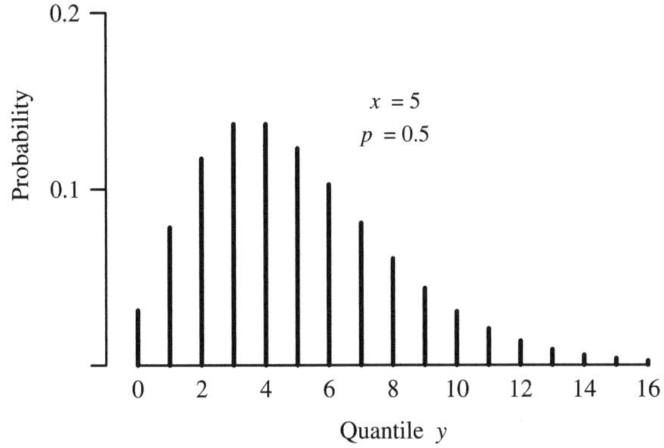

Figure 32.3. Probability function for the negative binomial variate ***NB***: *x, p*.

3. The sum of x-independent geometric variates $\boldsymbol{G}$: p is a negative binomial variate.

$$\sum_{i=1}^{x}(\boldsymbol{G} : p)_i \sim \boldsymbol{NB} : x, p$$

4. The negative binomial variate corresponds to the power series variate with parameter $c = 1 - p$, and probability function $(1 - c)^{-x}$.

5. As x tends to infinity and p tends to 1 with $x(1 - p) = \lambda$ held fixed, the negative binomial variate tends to the Poisson variate, $\boldsymbol{P} : \lambda$.

6. The binomial variate $\boldsymbol{B}$: $n, \ p$ and negative binomial variate $\boldsymbol{NB}$: $x, \ p$ are related by

$$\Pr\left[(\boldsymbol{B} : n, p) \leq x\right] = \Pr\left[(\boldsymbol{NB} : x, p) \geq (n - x)\right].$$

32.3 PARAMETER ESTIMATION

Parameter	Estimator	Method/Properties
p	$(x-1)/(y+x-1)$	Unbiased
p	$x/(y+x)$	Maximum likelihood

32.4 RANDOM NUMBER GENERATION

1. *Rejection Technique.* Select a sequence of unit rectangular random numbers, recording the numbers of those that are greater than and less than p. When the number less than p first reaches x, the number greater than p is a negative binomial random number, for x and y integer valued.

2. *Geometric Distribution Method.* If p is small, a faster method may be to add x geometric random numbers, as

$$\boldsymbol{NB} : x, p \sim \sum_{i=1}^{x}(\boldsymbol{G} : p)_i.$$

Chapter 33

Normal (Gaussian) Distribution

The normal distribution is applicable to a very wide range of phenomena and is the most widely used distribution in statistics. It was originally developed as an approximation to the binomial distribution when the number of trials is large and the Bernoulli probability p is not close to 0 or 1. It is also the asymptotic form of the sum of random variables under a wide range of conditions.

The normal distribution was first described by the French mathematician de Moivre in 1733. The development of the distribution is often ascribed to Gauss, who applied the theory to the movements of heavenly bodies.

Variate N: μ, σ.

Range $-\infty < x < \infty$.

Location parameter μ, the mean.

Scale parameter $\sigma > 0$, the standard deviation.

Probability density function	$\sigma(2\pi)^{1/2} \Big/ \exp\left(\dfrac{-(x-\mu)^2}{2\sigma^2}\right)$
Moment generating function	$\exp\left(\mu t + \dfrac{1}{2}\sigma^2 t^2\right)$
Characteristic function	$\exp\left(i\mu t - \dfrac{1}{2}\sigma^2 t^2\right)$
Cumulant generating function	$i\mu t - \dfrac{1}{2}\sigma^2 t^2$
rth Cumulant	$\kappa_1 = \mu, \kappa_2 = \sigma^2, \kappa_r = 0, \ r > 2$
Mean	μ

Statistical Distributions, Fourth Edition, by Catherine Forbes, Merran Evans, Nicholas Hastings, and Brian Peacock

rth Moment about the mean	$\begin{cases} \mu_r = 0 & r \text{ odd} \\ \mu_r = \sigma^r r!/\{2^{r/2}[(r/2)!]\} & \\ \quad = \sigma^r(r-1)(r-3)\cdots(3)(1) & r \text{ even} \end{cases}$
Variance	σ^2
Mean deviation	$\sigma(2/\pi)^{1/2}$
Mode	μ
Median	μ
Standardized rth moment about the mean	$\begin{cases} \eta_r = 0 & r \text{ odd} \\ \eta_r = r!/\{2^{r/2}[(r/2)!]\} & r \text{ even} \end{cases}$
Coefficient of skewness	0
Coefficient of kurtosis	3
Information content	$\log_2[\sigma(2\pi e)^{1/2}]$

The probability density function of the standard $N : 0, 1$ variate is shown in Figure 33.1, with the corresponding distribution and hazard functions shown in Figure 33.2 and 33.3, respectively.

33.1 VARIATE RELATIONSHIPS

The standard normal variate N: 0, 1 and the normal variate N: μ, σ are related by

$$N : 0, 1 \sim [(N : \mu, \sigma) - \mu]/\sigma.$$

1. Let $N_i, i = 1, \ldots, n$ be independent normal variates with means μ_i and variances σ_i^2. Then $\sum_{i=1}^n c_i N_i$ is normally distributed with mean $\sum_{i=1}^n c_i \mu_i$ and variance $\sum_{i=1}^n c_i^2 \sigma_i^2$, where the $c_i, i = 1, \ldots, n$ are constant weighting factors.

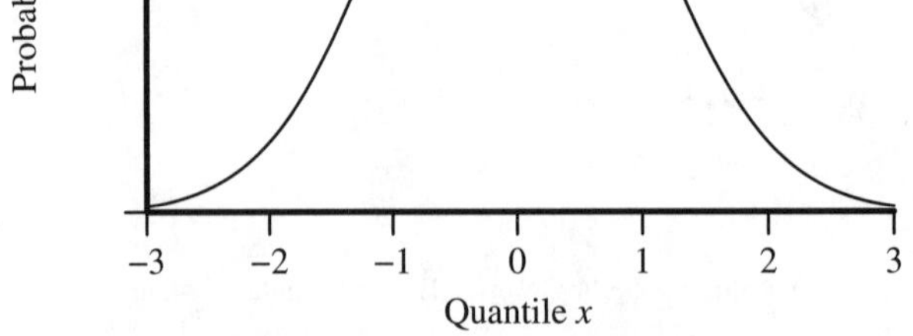

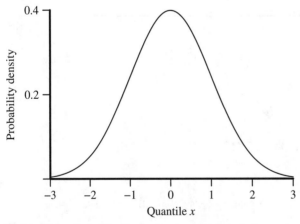

Figure 33.1. Probability density function for the standard normal variate N: 0, 1.

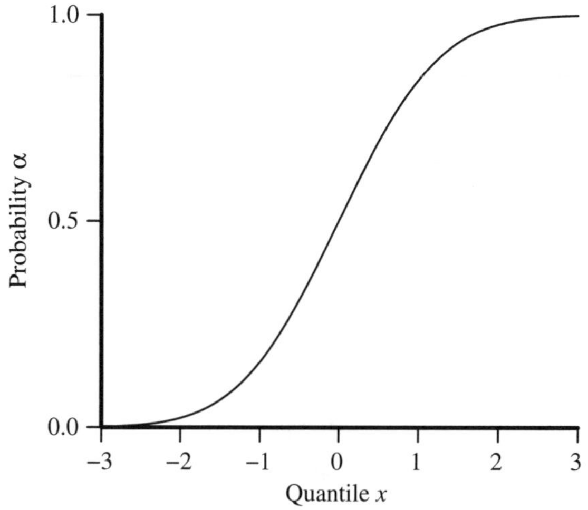

Figure 33.2. Distribution function for the standard normal variate *N*: 0, 1.

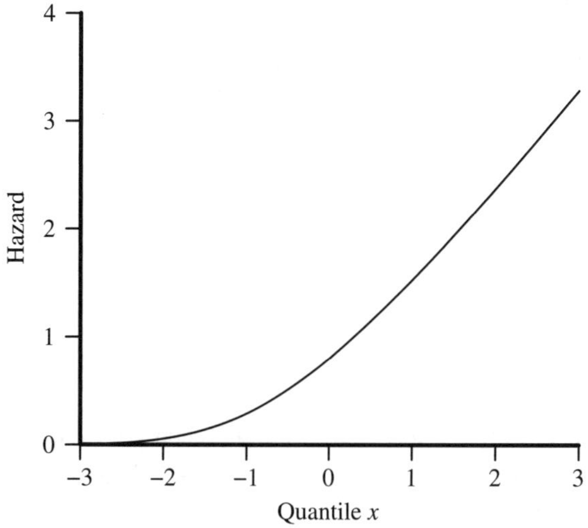

Figure 33.3. Hazard function for the standard normal variate *N*: 0, 1.

2. The sum of n-independent normal variates, $\boldsymbol{N} : \mu, \sigma$, is a normal variate with mean $n\mu$ and standard deviation $\sigma n^{1/2}$.

$$\sum_{i=1}^{n} (\boldsymbol{N} : \mu, \sigma)_i \sim \boldsymbol{N} : n\mu, \sigma n^{1/2}.$$

3. Any fixed linear transformation of a normal variate is also a normal variate. For constants a and b,

$$a + b(\boldsymbol{N} : \mu, \sigma) \sim \boldsymbol{N} : a + \mu, b\sigma.$$

4. The sum of the squares of ν-independent unit normal variates, $\boldsymbol{N}$: 0, 1, is a chi-squared variate with ν degrees of freedom, $\boldsymbol{\chi}^2 : \nu$:

$$\sum_{i=1}^{\nu}(N : 0, 1)_i^2 \sim \chi^2 : \nu$$

and for δ_i, $i = 1, \ldots, \nu$, and $\delta = \sum_{i=1}^{\nu} \delta_i^2$,

$$\sum_{i=1}^{\nu}[(N : 0, 1) + \delta_i]^2 \sim \sum_{i=1}^{\nu}(N : \delta_i, 1)^2 \sim \chi^2 : \nu, \delta$$

where $\chi^2 : \nu, \delta$ is the noncentral chi-squared variate with parameters ν, δ.

5. The normal variate $\boldsymbol{N} : \mu, \sigma$ and the lognormal variate $\boldsymbol{L} : \mu, \sigma$ are related by

$$L : \mu, \sigma \sim \exp(N : \mu, \sigma).$$

6. The ratio of two independent $\boldsymbol{N}$: 0, 1 variates is the standard Cauchy variate with parameters 0 and 1, here denoted $\boldsymbol{C}$: 0, 1,

$$C : 0, 1 \sim (N : 0, 1)_1/(N : 0, 1)_2.$$

7. The standardized forms of the following variates tend to the standard normal variate $\boldsymbol{N}$: 0, 1:

Binomial $\boldsymbol{B}$: n, p as n tends to infinity.

Beta $\boldsymbol{\beta} : \nu, \omega$ as ν and ω tend to infinity such that ν/ω is constant.

Chi-squared $\boldsymbol{\chi}^2 : \nu$ as ν tends to infinity.

Noncentral chi-squared $\boldsymbol{\chi}^2 : \nu, \delta$ as δ tends to infinity, such that ν remains constant, and also as ν tends to infinity such that δ remains constant.

Gamma $\boldsymbol{\gamma}$: b, c as c tends to infinity.

Inverse Gaussian $\boldsymbol{I} : \mu, \lambda$ as λ tends to infinity.

Lognormal $\boldsymbol{L} : \mu, \sigma$ as σ tends to zero.

Poisson $\boldsymbol{P} : \lambda$ as λ tends to infinity.

Student's $\boldsymbol{t} : \nu$ as ν tends to infinity.

8. The sample mean of n-independent and identically distributed random variates, each with mean μ and variance σ^2, tends to be normally distributed with mean μ and variance σ^2/n, as n tends to infinity.

9. If n-independent variates have finite means and variances, then the standardized form of their sample mean tends to be normally distributed, as n tends to infinity. These follow from the central limit theorem.

33.2 PARAMETER ESTIMATION

Parameter	Estimator	Method/Properties
μ	$\bar{x}$	Unbiased, maximum likelihood
σ^2	$ns^2/(n-1)$	Unbiased
σ^2	s^2	Maximum likelihood

33.3 RANDOM NUMBER GENERATION

Let $\boldsymbol{R}_1$ and $\boldsymbol{R}_2$ denote independent unit rectangular variates. Then two independent standard normal variates are generated by

$$\sqrt{-2\log \boldsymbol{R}_1}\sin(2\pi \boldsymbol{R}_2)$$
$$\sqrt{-2\log \boldsymbol{R}_1}\cos(2\pi \boldsymbol{R}_2).$$

33.4 TRUNCATED NORMAL DISTRIBUTION

When normal variates $\boldsymbol{N}: \mu,\ \sigma$ are restricted to lie within an interval $[a, b]$ the distribution of the variate is said to be truncated. The resulting characteristics of the so-called truncated normal variate $\boldsymbol{X}$ may be described in terms of the probability density function, $f_N(x)$, and the probability distribution function, $F_N(x)$, respectively, of the standard normal variate $\boldsymbol{N}: 0,\ 1$.

Variate $\boldsymbol{X}: \mu,\ \sigma,\ a,\ b$.

Range $a \leq x \leq b$.

Location parameter μ.

Scale parameter σ.

Trunctation parameters a (lower) and b (upper).

Let $a^* = \frac{a-\mu}{\sigma}$ and $b^* = \frac{b-\mu}{\sigma}$.

Probability density function $$\frac{f_N(x)}{[F_N(b^*) - F_N(a^*)]}$$

Moment generating function $$\exp\left(\mu t + \frac{\sigma^2 t^2}{2}\right)\frac{[F_N(b^* - t\sigma) - F_N(a^* - t\sigma)]}{[F_N(b^*) - F_N(a^*)]}$$

Characteristic function $$\exp\left(i\mu t - \frac{\sigma^2 t^2}{2}\right)\frac{F_N(b^* - it\sigma) - F_N(a^* - it\sigma)}{F_N(b^*) - F_N(a^*)}$$

Mean $$\mu + \sigma\frac{f_N(a^*) - f_N(b^*)}{[F_N(b^*) - F_N(a^*)]}$$

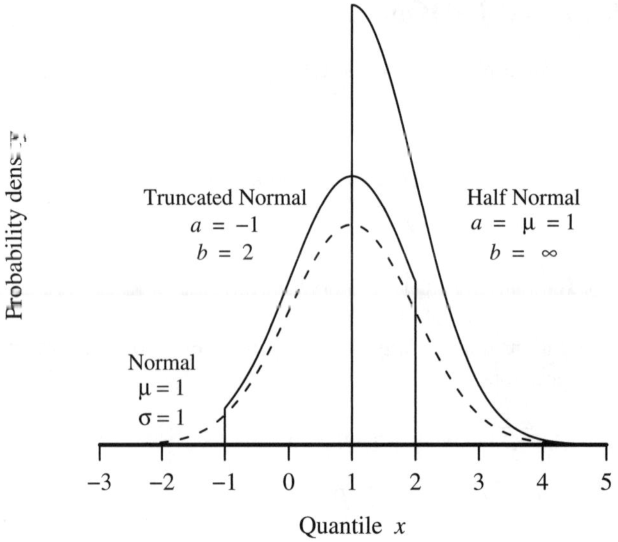

Figure 33.4. Probability density function for the truncated normal variate $X : \mu, \sigma, a, b$.

Variance $$\sigma^2\left[1+\frac{a^* f_N(a^*)-b^* f_N(b^*)}{[F_N(b^*)-F_N(a^*)]}-\left(\frac{[f_N(a^*)-f_N(b^*)]}{[F_N(b^*)-F_N(a^*)]}\right)^2\right]$$

The probability density functions for the truncated normal variate $X : 1, 1, -1, 2$ variate and the half normal variate $X : 1, 1, -2$, are both shown in Figure 33.4, along with the corresponding pdf for the $N : 1, 1$ variate.

33.5 VARIATE RELATIONSHIPS

1. The truncated normal variate $X : \mu, \sigma, a, b$ is the same as the normal variate N: μ, σ if both $a = -\infty$ and $b = \infty$.

2. If for the truncated normal variate $X : \mu, \sigma, a, b$ either $a = \mu$ and $b = \infty$, or $a = -\infty$ and $b = \mu$, then the variate is called a half normal variate.

Chapter 34

Pareto Distribution

The Pareto distribution is often described as the basis of the 80/20 rule. For example, 80% of customer complaints regarding a make of vehicle typically arise from 20% of components. Other applications include the distribution of income and the classification of stock in a warehouse on the basis of frequency of movement.

Range $a \leq x < \infty$.

Location parameter $a > 0$.

Shape parameter $c > 0$.

Distribution function	$1 - (a/x)^c$
Probability density function	ca^c/x^{c+1}
Inverse distribution function (of probability α)	$a(1-\alpha)^{-1/c}$
Survival function	$(a/x)^c$
Inverse survival function (of probability α)	$a\alpha^{-1/c}$
Hazard function	c/x
Cumulative hazard function	$c \log (x/a)$
rth Moment about the mean	$ca^r/(c-r), \quad c > r$
Mean	$ca/(c-1), \quad c > 1$
Variance	$ca^2/[(c-1)^2(c-2)], \quad c > 2$
Mode	a
Median	$2^{1/c}a$
Coefficient of variation	$[c(c-2)]^{-1/2}, \quad c > 2$

34.1 NOTE

This is a Pareto distribution of the first of three kinds. Stable Pareto distributions have $0 < c < 2$.

Statistical Distributions, Fourth Edition, by Catherine Forbes, Merran Evans, Nicholas Hastings, and Brian Peacock

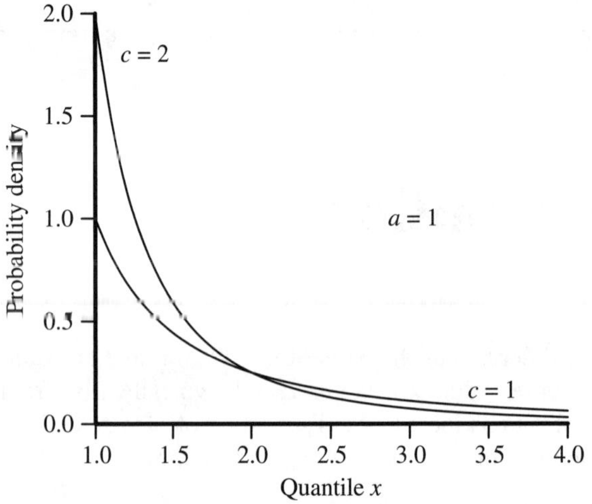

Figure 34.1. Probability density function for the Pareto variate X: a, c.

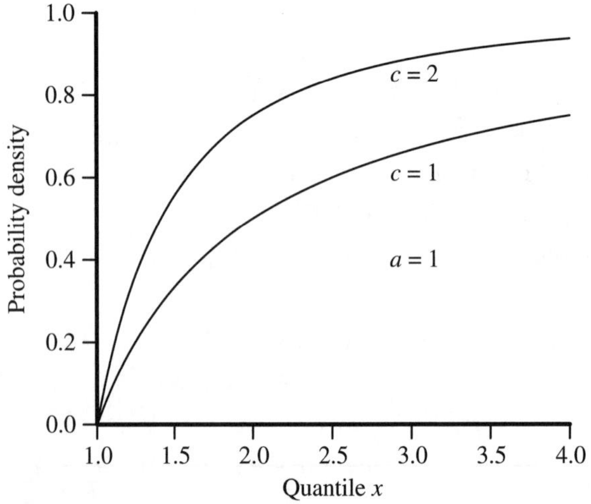

Figure 34.2. Distribution function for the Pareto variate X: a, c.

The probability density function for the Pareto variate $X : 1, c$ is shown in Figure 34.1 for selected values of the shape parameter, c. The corresponding distribution functions are shown in Figure 34.2.

34.2 VARIATE RELATIONSHIPS

1. The Pareto variate, here denoted X: a, c, is related to the following variates:

   The exponential variate E: b with parameter $b = 1/c$,

$$\log[(X : a, c)/a] \sim E : 1/c.$$

The power function variate $\boldsymbol{Y}$: b, c with parameter $b = 1/a$,

$$[\boldsymbol{X} : a, c]^{-1} \sim \boldsymbol{Y} : 1/a, c.$$

The standard logistic variate, here denoted $\boldsymbol{Y}$: 0, 1,

$$-\log\{[(\boldsymbol{X} : a, c)/a]^{c} - 1\} \sim \boldsymbol{Y} : 0, 1.$$

2. The n-independent Pareto variates, $\boldsymbol{X}$: a, c, are related to a standard gamma variate with shape parameter n, $\boldsymbol{\gamma}$: 1, n, and to a chi-squared variate with $2n$ degrees of freedom by

$$\begin{aligned} 2a\sum_{i=1}^{n}\log[(\boldsymbol{X} : a, c)_i/c] &= 2a\log\prod_{i=1}^{n}(\boldsymbol{X} : a, c)_i/c^n \\ &\sim \boldsymbol{\gamma} : 1, n \\ &\sim \chi^2 : 2n. \end{aligned}$$

34.3 PARAMETER ESTIMATION

Parameter	Estimator	Method/Properties
$1/c$	$\left(\frac{1}{n}\right)\sum_{i=1}^{n}\log\left(\frac{x_i}{\hat{a}}\right)$	Maximum likelihood
a	$\min x_i$	Maximum likelihood

34.4 RANDOM NUMBER GENERATION

The Pareto variate $\boldsymbol{X}$: a, c is related to the unit rectangular variate $\boldsymbol{R}$ by

$$\boldsymbol{X} : a, c, \sim a(1 - \boldsymbol{R})^{-1/c}.$$

Chapter 35

Poisson Distribution

The Poisson distribution is applied in counting the number of rare but open-ended events. A classic example is the number of people per year who become invalids due to being kicked by horses. Another application is the number of faults in a batch of materials.

It is also used to represent the number of arrivals, say, per hour, at a service center. This number will have a Poisson distribution if the average arrival rate does not vary through time. If the interarrival times are exponentially distributed, the number of arrivals in a unit time interval are Poisson distributed. In practice, arrival rates may vary according to the time of day or year, but a Poisson model will be used for periods that are reasonably homogeneous.

The mean and variance are equal and can be estimated by observing the characteristics of actual samples of "arrivals" or "faults." See Chapter 38.

Variate $\boldsymbol{P} : \lambda$.

Range $0 \leq x < \infty$, x an integer.

Parameter $\lambda > 0$, the mean, also the variance.

Distribution function	$\sum_{i=1}^{X} \lambda^i \exp(-\lambda)/i!$
Probability function	$\lambda^x \exp(-\lambda)/x!$
Moment generating function	$\exp\{\lambda[\exp(t) - 1]\}$
Probability generating function	$\exp[\lambda(t - 1)]$
Characteristic function	$\exp\{\lambda[\exp(it) - 1]\}$
Cumulant generating function	$\lambda[\exp(it) - 1] = \lambda \sum_{j=1}^{\infty} (it)^j/j!$
rth Cumulant	λ
Moments about the origin	
Mean	λ
Second	$\lambda + \lambda^2$

Statistical Distributions, Fourth Edition, by Catherine Forbes, Merran Evans, Nicholas Hastings, and Brian Peacock

Third	$\lambda[(\lambda+1)^2+\lambda]$
Fourth	$\lambda(\lambda^3+6\lambda^2+7\lambda+1)$
rth Moment about the mean	$\lambda\sum_{i=0}^{r-2}\binom{r-1}{i}\mu_i, \quad r>1, \quad \mu_0=1$
Moments about the mean	
Variance	λ
Third	λ
Fourth	$\lambda(1+3\lambda)$
Fifth	$\lambda(1+10\lambda)$
Sixth	$\lambda(1+25\lambda+15\lambda^2)$
Mode	The mode occurs when x is the largest integer less than λ. For λ an integer the values $x=\lambda$ and $x=\lambda-1$ are tie modes.
Coefficient of skewness	$\lambda^{-1/2}$
Coefficient of kurtosis	$3+1/\lambda$
Coefficient of variation	$\lambda^{-1/2}$
Factorial moments about the mean	
Second	λ
Third	-2λ
Fourth	$3\lambda(\lambda+2)$

35.1 NOTE

Successive values of the probability function $f(x)$, for $x = 0, 1, 2, \ldots$, are related by

$$f(x+1) = \lambda f(x)/(x+1)$$
$$f(0) = \exp(-\lambda).$$

The probability function for the $\boldsymbol{P}:\lambda$ variate is shown in the panels of Figure 35.1 for selected values of λ, with the corresponding distribution functions shown in the panels of Figure 34.2.

35.2 VARIATE RELATIONSHIPS

1. The sum of a finite number of independent Poisson variates, $\boldsymbol{P}:\lambda_1, \boldsymbol{P}:\lambda_2, \ldots, \boldsymbol{P}:\lambda_n$, is a Poisson variate with mean equal to the sum of the means of

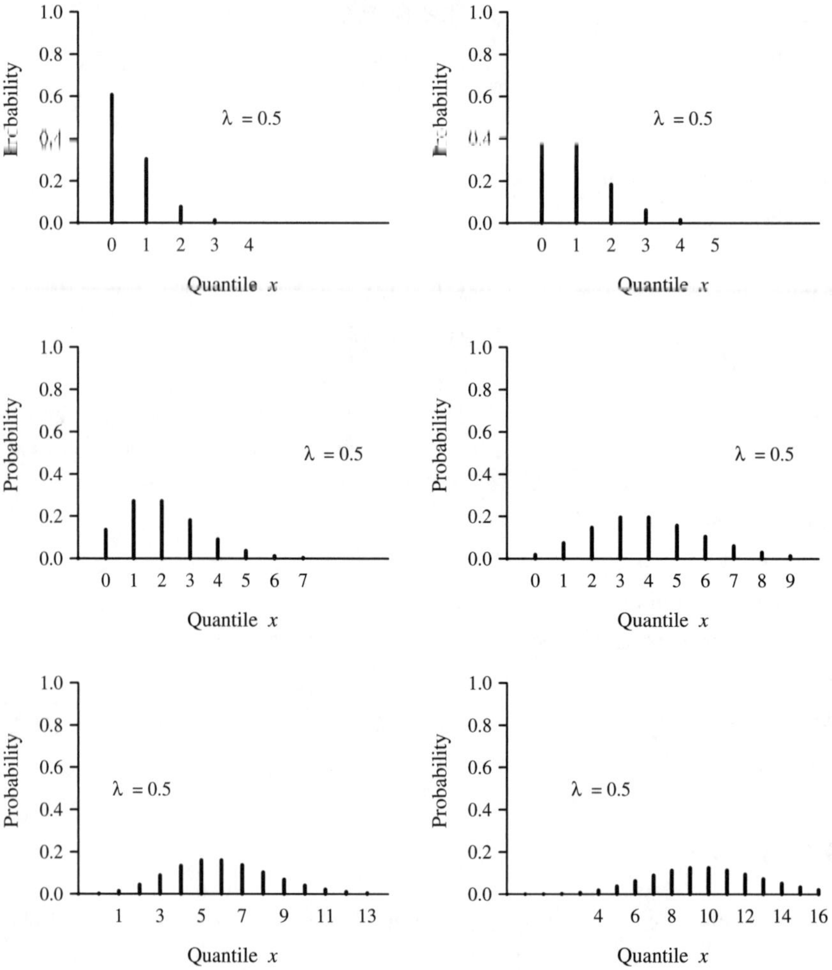

Figure 35.1. Probability function for the Poisson variate $\boldsymbol{P}: \lambda$.

the separate variates:

$$(\boldsymbol{P}:\lambda_1)+(\boldsymbol{P}:\lambda_2)+\cdots+(\boldsymbol{P}:\lambda_n)$$
$$\sim(\boldsymbol{P}:\lambda_1+\lambda_2+\cdots+\lambda_n).$$

2. The Poisson variate $\boldsymbol{P}:\lambda$ is the limiting form of the binomial variate $\boldsymbol{B}$: n, p, as n tends to infinity and p tends to zero such that np tends to λ.

$$\lim_{n\to\infty, np\to\lambda}\left[\binom{n}{x}p^x(1-p)^{n-x}\right]=\lambda^x\exp(-\lambda)/x!$$

3. For large values of λ the Poisson variate $\boldsymbol{P}:\lambda$ may be approximated by the normal variate with mean λ and variance λ.

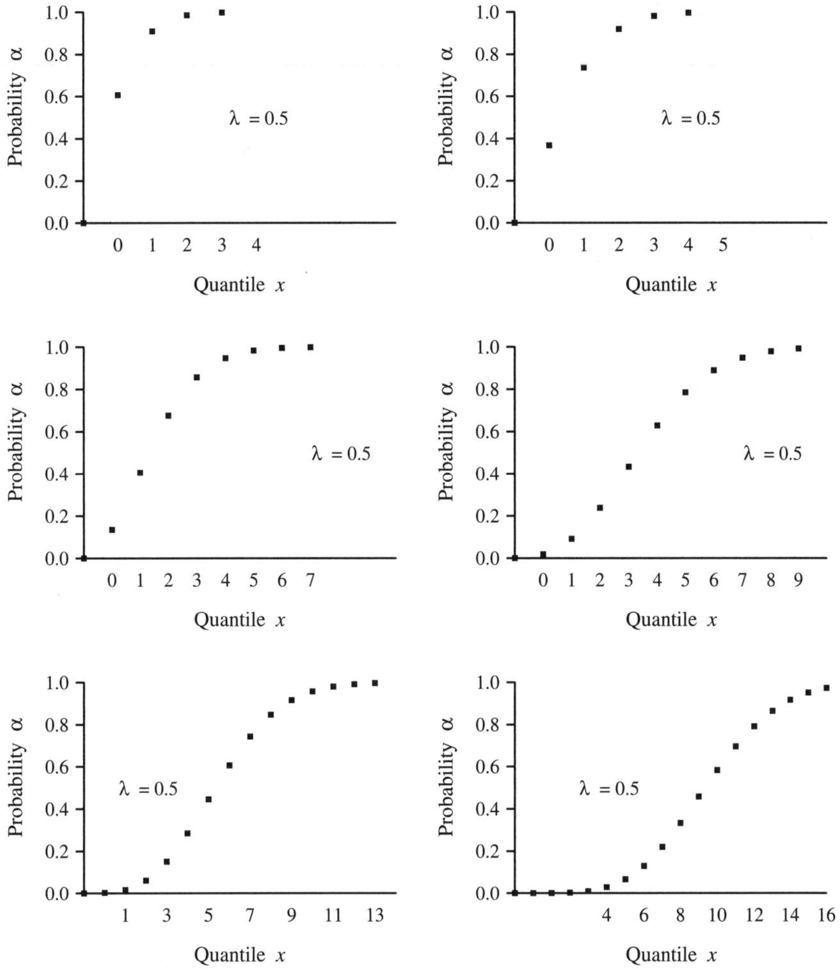

Figure 35.2. Distribution function for the Poisson variate $\boldsymbol{P}:\lambda$.

4. The probability that the Poisson variate $\boldsymbol{P}:\lambda$ is less than or equal to x is equal to the probability that the chi-squared variate with $2(1+x)$ degrees of freedom, denoted $\chi^2 : 2(1+x)$, is greater than 2λ.

$$\Pr[(\boldsymbol{P}:\lambda) \leq x] = \Pr[(\chi^2 : 2(1+x)) > 2\lambda]$$

5. The hypergeometric variate $\boldsymbol{H}$: N, X, n tends to a Poisson variate $\boldsymbol{P}:\lambda$ as X, N, and n all tend to infinity, for X/N tending to zero, and nX/N tending to λ.

6. The Poisson variate $\boldsymbol{P}:\lambda$ is the power series variate with parameter λ and series function $\exp(\lambda)$.

35.3 PARAMETER ESTIMATION

Parameter	Estimator	Method/Properties
λ	$\bar{x}$	Maximum likelihood, Minimum variance unbiased

35.4 RANDOM NUMBER GENERATION

Calculate the distribution function $F(x)$ for $x = 0, 1, 2, \ldots, N$, where N is an arbitrary (large) cut off number. Choose random numbers of the unit rectangular variate $\boldsymbol{R}$. If $F(x) \leq \boldsymbol{R} < F(x + 1)$, then the corresponding Poisson random number is x.

Chapter 36

Power Function Distribution

Range $0 \leq x \leq b$.

Shape parameter c, scale parameter $b > 0$.

Distribution function	$(x/b)^c$
Probability density function	cx^{c-1}/b^c
Inverse distribution function (of probability α)	$b\alpha^{1/c}$
Hazard function	$cx^{c-1}/(b^c - x^c)$
Cumulative hazard function	$-\log[1-(x/b)^c]$
rth Moment about the origin	$b^r c/(c+r)$
Mean	$bc/(c+1)$
Variance	$b^2c/[(c+2)(c+1)^2]$
Mode	b for $c > 1, 0$ for $c < 1$
Median	$b/2^{1/c}$
Coefficient of skewness	$2(1-c)(2+c)^{1/2}/((3+c)c^{1/2})$
Coefficient of kurtosis	$3(c+2)[2(c+1)^2+c(c+5)]/[c(c+3)(c+4)]$
Coefficient of variation	$[c(c+2)]^{-1/2}$

The probability density function for the power function variate $X : 1, c$ is shown in Figure 36.1, for selected values of the shape parameter c, with the corresponding distribution function is shown in Figure 36.2.

36.1 VARIATE RELATIONSHIPS

1. The power function variate with scale parameter b and shape parameter c, here denoted X: b, c, is related to the power function variate X: $1/b$, c by

$$[X : b, c]^{-1} \sim X : \frac{1}{b}, c.$$

Statistical Distributions, Fourth Edition, by Catherine Forbes, Merran Evans, Nicholas Hastings, and Brian Peacock

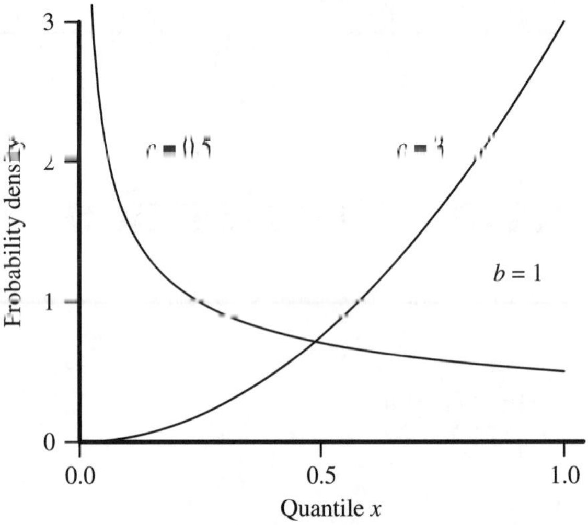

Figure 36.1. Probability density function for the power function variate **X** : b, c.

2. The standard power function variate, denoted X: 1, c, is a special case of the beta variate, $\boldsymbol{\beta} : \nu, \omega$, with $\nu = c$, $\omega = 1$.

$$X : 1, c \sim \boldsymbol{\beta} : c, 1$$

3. The standard power function, denoted X: 1, c, is related to the following variates:

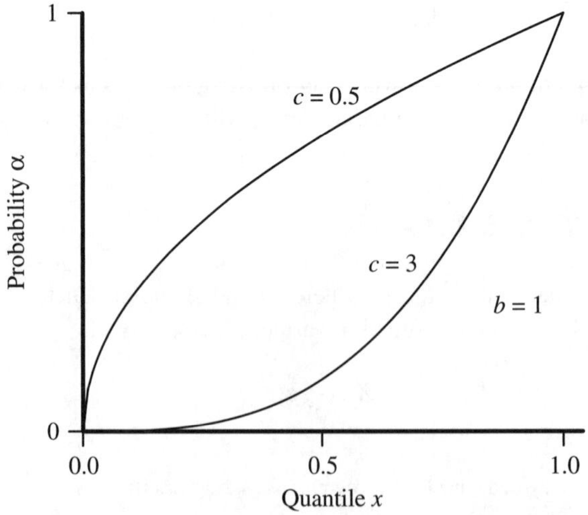

Figure 36.2. Distribution function for the power function variate **X** : b, c.

The exponential variate $\boldsymbol{E}$: b with shape parameter $b = 1/c$,

$$-\log[\boldsymbol{X} : 1, c] \sim \boldsymbol{E} : 1/c.$$

The Pareto variate with location parameter zero and shape parameter c, here denoted $\boldsymbol{Y}$: 0, c,

$$[\boldsymbol{X} : 1, c]^{-1} \sim \boldsymbol{Y} : 0, c.$$

The standard logistic variate, here denoted $\boldsymbol{Y}$: 0, 1,

$$-\log\{(\boldsymbol{X} : 1, c)^{-c} - 1\} \sim \boldsymbol{Y} : 0, 1.$$

The standard Weibull variate, with shape parameter k,

$$[-\log(\boldsymbol{X} : 1, c)^{c}]^{1/k} \sim \boldsymbol{W} : 1, k.$$

The standard Gumbel extreme value variate, $\boldsymbol{V}$: 0, 1,

$$-\log[-c\log(\boldsymbol{X} : 1, c)] \sim \boldsymbol{V} : 0, 1.$$

4. The power function variate with shape parameter $c = 1$, denoted $\boldsymbol{X}$: b, 1, corresponds to the rectangular variate $\boldsymbol{R}$: 0, b.

5. Two independent standard power function variates, denoted $\boldsymbol{X}$: 1, c, are related to the standard Laplace variate $\boldsymbol{L}$: 0, 1 by

$$-c\log[(\boldsymbol{X} : 1, c)_1/(\boldsymbol{X} : 1, c)_2] \sim \boldsymbol{L} : 0, 1.$$

36.2 PARAMETER ESTIMATION

Parameter	Estimator	Method/Properties
c	$n \Big/ \left[\sum_{j=1}^{n} \log x_j\right]$	Maximum likelihood
c	$\bar{x}/(1-\bar{x})$	Matching moments

36.3 RANDOM NUMBER GENERATION

The power function random variate $\boldsymbol{X}$: b, c can be obtained from the unit rectangular variate $\boldsymbol{R}$ by

$$\boldsymbol{X} : b, c \sim b(\boldsymbol{R})^{1/c}.$$

Chapter 37

Power Series (Discrete) Distribution

Range of x is a countable set of integers for generalized power series distributions.

Parameter $c > 0$.

Coefficient function $a_x > 0$, series function $A(c) = \Sigma a_x c^x$.

Probability function	$a_x c^x / A(c)$
Probability generating function	$A(ct)/A(c)$
Moment generating function	$A[c\exp(t)]/A(c)$
Mean, μ_1	$c\dfrac{d}{dc}[\log A(c)]$
Variance, μ_2	$\mu_1 + c^2\dfrac{d^2}{dc^2}[\log A(c)]$
rth Moment about the mean	$c\dfrac{d\mu_r}{dc} + r\mu_3\mu_{r-1}, \quad r > 2$
First cumulant, κ_1	$c\dfrac{d}{dc}[\log A(c)] = \dfrac{c}{A(c)}\dfrac{dA(c)}{dc}$
rth Cumulant, κ_r	$c\dfrac{d}{dc}\kappa_{r-1}$

37.1 NOTE

Power series distributions (PSDs) can be extended to the multivariate case. Factorial series distributions are the analogue of power series distributions, for a discrete parameter c. (See Kotz and Johnson, 1986, Vol. 7, p. 130.)

Generalized hypergeometric (series) distributions are a subclass of power series distributions.

Statistical Distributions, Fourth Edition, by Catherine Forbes, Merran Evans, Nicholas Hastings, and Brian Peacock

37.2 VARIATE RELATIONSHIPS

1. The binomial variate ***B***: n, p is a PSD variate with parameter $c = p/(1-p)$ and series function $A(c) = (1+c)^n = (1-p)^{-n}$.
2. The Poisson variate ***P***: λ is a PSD variate with parameter $c = \lambda$ and series function $A(c) = \exp(c)$ and is uniquely characterized by having equal mean and variance for any c.
3. The negative binomial variate ***NB***: x, p is a PSD variate with parameter $c = 1 - p$ and series function $A(c) = (1-c)^{-x} = p^{-x}$.
4. The logarithmic series variate is a PSD variate with parameter c and series function $A(c) = -\log(1-c)$.

37.3 PARAMETER ESTIMATION

The estimator $\hat{c}$ of the shape parameter, obtained by the methods of maximum likelihood or matching moments, is the solution of the equation

$$\bar{x} = \hat{c}\frac{d}{d\hat{c}}[\log A(\hat{c})] = \frac{\hat{c}}{A(\hat{c})}\frac{d[A(\hat{c})]}{d\hat{c}}.$$

Chapter 38

Queuing Formulas

A queuing or waiting line system is where customers arrive for service, possibly wait in line, are served, and depart. Examples are supermarket checkouts, airline check-ins, bank tellers, and so on. Queuing theory provides a classification system and mathematical analysis of basic queuing models and this assists in the conceptual understanding, design, and operation of queuing systems. Simulation techniques provide a way to create more elaborate models. Here we summarize the formulas for a number of the standard mathematical queuing models. Figure 38.1 illustrates the basic situation of customers arriving, possibly waiting in a queue, being served, and departing.

38.1 CHARACTERISTICS OF QUEUING SYSTEMS AND KENDALL-LEE NOTATION

We can describe various types of queuing system in terms of six characteristics. In the following sections we introduce these characteristics, the Kendall-Lee notation, which is used to describe them, and some terms and symbols.

Characteristics of Queuing Systems

Arrival Pattern

Inter-arrival times at a queuing system follow a statistical distribution, commonly taken to be a (negative) exponential distribution. The parameter λ (lambda) is the average arrival rate per unit time. An exponential arrival pattern is denoted by the letter M, for Markov, in the Kendall-Lee notation. The term *Poisson arrivals* is also used to describe this pattern, as the number of arrivals in any given time interval has a Poisson distribution.

Statistical Distributions, Fourth Edition, by Catherine Forbes, Merran Evans, Nicholas Hastings, and Brian Peacock

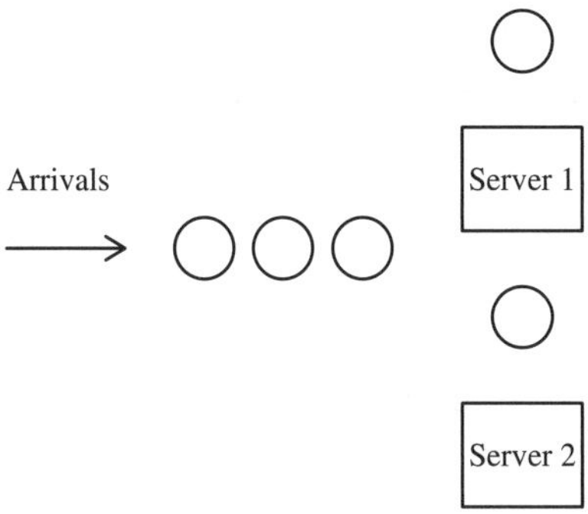

Figure 38.1. A queuing system.

Service Time

The servers, assumed to be identical, have service times that follow a statistical distribution. If the distribution is exponential, the average service rate per server is denoted by μ. The mean service time per server for any service time distribution is denoted by τ. In the Kendall-Lee notation, an exponential service pattern is denoted by the letter M, a general service-time distribution is denoted by G, and a deterministic service time is denoted by D.

Number of Servers

The number of servers is denoted by s.

Queue Discipline

The queue discipline is the rule for taking items from the queue. The most common is *first come first served*, denoted by $FCFS$. Many mathematical results do not depend on the queue discipline and apply for a general discipline, denoted by G.

Maximum Number in System

The maximum number of customers in the system is denoted by C. The default value is infinity, that is, no limit on the number of customers in the system.

Population

The size of the population from which the customers are drawn is denoted by N. The default value is infinity, that is, no limit on the population from which the customers are drawn.

Customer Behavior

The following features of customer behavior sometimes occur. Analysis of queues involving these behaviors is carried out by simulation rather than by application of mathematical formulas.

- *Balking* is deciding not to join a queue dependent on its length;
- *Reneging* is leaving a queue after you have joined;
- *Jockeying* is switching from queue to queue in multiple queue systems.

Kendall-Lee Notation

Kendall-Lee notation is a shorthand way of describing a queuing system based on the characteristics just described. The value of each characteristic is denoted by its letter symbol or a suitable numeric value, for example:

$$M/M/2/FCFS/6/\infty$$

indicates a queuing system with Poisson arrivals, exponential service, two servers, first come first served ($FCFS$) queue discipline, a limit of six customers in the queue and an unlimited source of customers in the population.

38.2 DEFINITIONS, NOTATION, AND TERMINOLOGY

Steady State

Queuing formulas apply to steady state average values of properties such as the average length of the queue, the average time a customer spends in the queuing system and so on. Systems reach a steady state only if the arrival rate is less than the maximum service rate, or if excess arrivals are rejected by the system due to limits on the length of the queue.

Traffic Intensity and Traffic Density

For the basic $M/M/1$ queue the traffic intensity is the ratio of the arrival rate to the service rate, and this must be less than one for the system to reach a steady state. For other systems, the ratio of an arrival rate to a service rate appears in many queuing formulas. We refer to this quantity as the traffic density. The traffic density can be more than one and the system will reach equilibrium if there are sufficient servers or if the there are queue restrictions. Figure 38.2 demonstrates the average number in an $M/M/1$ system as a function of the traffic intensity.

Notation and Terminology

Table 38.1 displays the symbols associated with the many varied queuing systems described in this chapter.

Table 38.1: Queuing Notation

Symbol	Description
λ	Customer arrival rate
μ	Service rate per server for exponential service
τ	Mean service time for general distribution ($\tau = 1/\mu$ for exponential service)
s	Number of servers
ρ	Traffic intensity (λ/μ for $M/M/1$ queue) (Check definition for other systems.)
d	Traffic density (λ/μ or $\lambda\tau$)
L	Mean number of customers in the system
m	Maximum number of customers in the system when this is limited
λ_a	In-system customer arrival rate when some customers do not join the system
L_q	Mean number of customers in the queue (waiting line)
L_s	Mean number of customers in service
W	Mean time a customer spends in the system
W_q	Mean time a customer spends in the queue
W_s	Mean time a customer spends in service
P_j	Steady state probability that there are exactly j customers in the system
N	Population in limited population systems

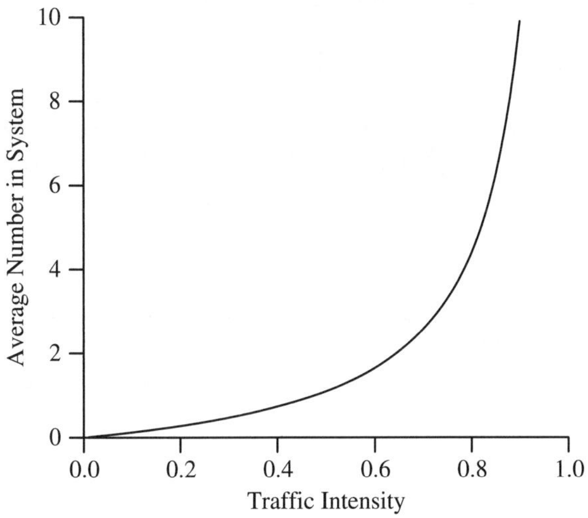

Figure 38.2. $M/M/1$ system. Average number of customers in system, L, versus traffic intensity ρ.

38.3 GENERAL FORMULAS

Little's formula,

$$L = \lambda W$$

is based on equilibrium considerations and applies to a wide range of queuing systems. The formula simply asserts that the average number of customers in the system is equal to the customer arrival rate multiplied by the average time a customer spends in the system. Other general formulas are

$$\begin{aligned} L_q &= \lambda W_q \\ W_s &= \tau \\ L_s &= \lambda W_s = \lambda \tau \\ W &= W_q + W_s = W_q + \tau \\ L &= L_q + L_s = L_q + \lambda \tau. \end{aligned}$$

The delay ratio, denoted by H, is equal to the average time in the queue divided by average time in service, with

$$H = W_q/W_s = (L/\lambda\tau) - 1.$$

For equilibrium the maximum service rate must exceed the arrival rate

$$s > \lambda\tau$$

or, if some customers do not join, so that the in-system arrival rate is $\lambda_a \leq \lambda$, then

$$s > \lambda_a \tau.$$

38.4 SOME STANDARD QUEUING SYSTEMS

The *M/M/1/G/∞/∞* System

This model has Poisson arrivals at rate λ, exponential service at rate μ, one server, and no limit on the queue length or the population. The formulas for this system are shown in Table 38.2.

Application

The practical result from this model is that for the queue to be kept at moderate length the traffic intensity should not exceed 0.75, that is, the arrival rate must not be more than 75% of the service rate, and the server should be idle (or relaxed) 25% of the time.

The *M/M/s/G/ ∞/∞* System

This model has Poisson arrivals at rate λ, exponential service at rate μ, s servers and no limit on the queue length or the population. The formulas for this system are shown in Table 38.3.

Table 38.2: $M/M/1/G/\infty/\infty$ System Formulas

Description	Symbol	Formula
Traffic intensity	ρ	λ/μ
Probability server is idle	P_0	$1-\rho$
Probability j customers in system	P_j	$\rho^j(1-\rho)$
Probability server is busy	ρ	
Average number in system	L	$\rho/(1-\rho)=\lambda/(\mu-\lambda)$
Average number in queue	L_q	$\lambda^2/[\mu(\mu-\lambda)]=\rho^2/(1-\rho)$
Average number in service	L_s	ρ
Average time in system	W	$1/(\mu-\lambda)$
Average time in queue	W_q	$\rho/[\mu(1-\rho)]$
Average time in service	W_s	$1/\mu$
Delay ratio	H	$W_q/W_s=\rho/(1-\rho)$

Table 38.3: $M/M/s/G/\infty/\infty$ System Formulas

Description	Symbol	Formula
Traffic intensity	ρ	$\lambda/s\mu$
Traffic density	d	λ/μ
Probability all servers idle	P_0	$1\Big/\left[\sum_{i=0}^{s-1}(s\rho)^i/i!+(s\rho)^s/s!(1-\rho)\right]$
Probability j customers in system	P_j	$P_0(s\rho)^j/j!$ for $j=1,2,\ldots,s$ $P_0(s\rho)^j/\left[s!s^{j-s}\right]$ for $j>s$
Probability all servers busy	$P_{j\geq s}$	$P_0(s\rho)^s/[s!(1-\rho)]$
Proportion of servers busy	ρ	
Average number in queue	L_q	$P_{j\geq s}/(1-\rho)$
Average number in service	L_s	$\lambda/\mu=d$
Average number in system	L	$L_q+\lambda/\mu$
Average time in system	W	L/λ
Average time in queue	W_q	$P_{j\geq s}/(s\mu-\lambda)$
Average time in service	W_s	$1/\mu$
Delay ratio	H	$W_q/W_s=\mu P_{j\geq s}/(s\mu-\lambda)$
For FCFS service discipline		
Probability time in queue $>t$		$P_{j\geq s}\exp[-s\mu(1-\rho)t]$

Table 38.4: $M/G/1/G/\infty/\infty$ System Formulas

Description	Symbol	Formula
Coefficient of variation of service time distribution	v	σ/τ
Traffic intensity	ρ	$\lambda\tau$
Average number in queue	L_q	$\left(\lambda^2\sigma^2 + \rho^2\right)/[2(1-\rho)]$
If the service time is deterministic, it's standard deviation is zero		
Average number in queue, deterministic server	$L_q(D)$	$\rho^2/[2(1-\rho)]$
Ratio average number in queue for server with coefficient of variation, v to deterministic server	$L_q/L_q(D)$	$v^2 + 1$
Probability server is idle	P_0	$1 - \rho$
Probability server is busy	ρ	
Average number in system	L	$L_q + \rho$
Average number in service	L_s	ρ
Average time in queue	W_q	L_q/λ
Average time in system	W	$W_q + 1/\mu$
Average time in service	W_s	τ
Delay ratio	H	$W_q/W_s = L_q/(\lambda\tau)$
Delay ratio relative to deterministic server	H	$v^2 + 1$

The $M/G/1/G/\infty/\infty$ System (Pollaczek-Khinchin)

This model has Poisson arrivals at rate λ, a general service distribution with mean τ and standard deviation σ, one server, and no limit on the queue length or the population. The results for this queue were derived by Pollaczek and Khinchin. The formulas for this system are shown in Table 38.4.

Application

The practical result is that the queue length increases with the square of the coefficient of variation of the service time distribution. Hence the length of the queue in front of a service facility will be reduced if the variability of the service time is reduced. Figure 38.3 displays the average length of the queue under this system as a function of the coefficient of variation of service time.

The $M/M/1/G/m/\infty$ System

This model has Poisson arrivals at rate λ, exponential service rate μ, one server, a limit of m on the number of customers in the system and no limit on the population.

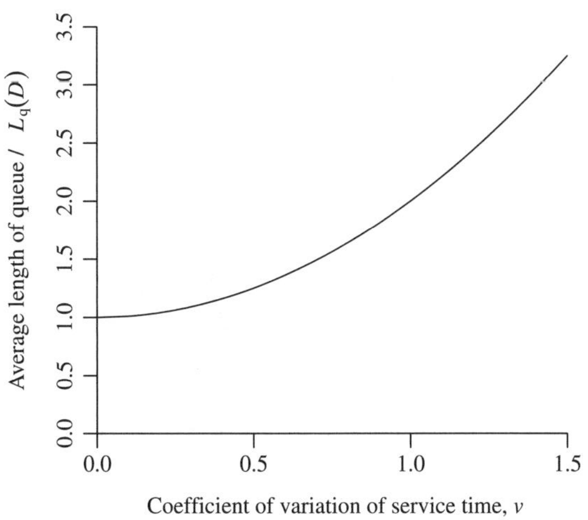

Figure 38.3. $M/G/1/G/\infty/\infty$ system. Queue length as a function of the coefficient of variation of the service time distribution, as a multiple of the length for a deterministic server.

Arrivals who find that there are m customers in the system (or equivalently, $m - 1$ in the queue) do not enter the system. Equilibrium can be achieved even with λ greater than or equal to μ, because arrivals who find the queue too long do not join the system. The formulas for this system are shown in Table 38.5.

The *M/G/m/G/m/∞* System (Erlang)

This model has Poisson arrivals at rate λ, general service distribution with mean τ, m servers, a limit of m on the number in the system and no limit on the population. Arrivals who find that there are m customers in the system, that is, all servers are busy, do not enter the system. Equilibrium can be achieved even with λ greater than or equal to μ, because arrivals who find the queue too long do not join the system. The formulas for this system are shown in Table 38.6.

Application

This system was analyzed by Erlang for application to a telephone system where m service channels were provided and customers who found them all busy were assumed to be lost to the system. The probability that all servers are busy is Erlang's loss formula. The analysis is used to choose a value of m, the number of servers, so that the loss of customers is at an acceptable level.

The *M/M/1/G/N/N* System (One Server, Finite Population *N*)

This model has a finite population, N. A common application is where the items are machines and so we will refer to the items in the population as machines. Each

Table 38.5: $M/M/1/G/m/\infty$ System Formulas

Description	Symbol	Formula
Traffic intensity	ρ	λ/μ
Maximum queue length	$m - 1$	
For $\lambda \neq \mu$		
Probability server is idle	P_0	$(1-\rho)/(1-\rho^{m+1})$
Probability j customers in the system	P_j	$P_0\rho^j,\ j = 1, 2, \ldots, m$
Probability system is full	P_m	$P_0\rho^m$
Average number in system	L	$p[1-(m+1)p^m + mp^{m+1}]/$ $[(1-p^{m+1})(1-p)]$
For $\lambda = \mu$		
Probability server is idle	P_0	$1/(m+1)$
Probability j customers in the system	P_j	$1/(m+1)$
Probability system is full	P_m	$1/(m+1)$
Average number in system	L	$m/2$
For all		
Proportion of arrivals lost to system	P_m	
Accepted customer rate		$\lambda(1-P_m)$
Average number in service	L_s	$1-P_0$
Average number in queue	L_q	$L - L_s$
Average time in system	W	$L/[\lambda(1-P_m)]$
Average time in queue	W_q	$L_q/[\lambda(1-P_m)]$
Average time in service	W_s	$1/\mu$
Delay ratio	H	$W_q/W_s = L_q/[\rho(1-P_m)]$

Table 38.6: $M/G/m/G/m/\infty$ System Formulas (Erlang)

Description	Symbol	Formula
Probability all servers are idle	P_0	$1/\left[\sum_{k=0}^{m}(\lambda\tau)^k/k!\right]$
Probability j servers are busy		$P_0(\lambda\tau)^j/j!,\ j = 0, \ldots, m$
Probability all servers are busy	P_m	$\left[(\lambda\tau)^m/m!\right] / \left[\sum_{k=0}^{m}(\lambda\tau)^k/k!\right]$
Proportion of arrivals lost to the system	P_m	
Offered load		$\lambda\tau$
Carried load		$\lambda\tau(1-P_m)$
Lost load		$\lambda\tau P_m$

Table 38.7: $M/M/1/G/N/N$ System Formulas

Description	Symbol	Formula
Traffic density	d	λ/μ
Probability the server is idle	P_0	$1/\left[\sum_{n=0}^{N} d^n N!/(N-n)!\right]$
Probability the server is busy	L_s	$1 - P_0$
Probability n items in the system	P_n	$P_0[d^n N!/(N-n)!]$ for $1 \leq n \leq N$
Average number in queue	L_q	$N - [(1+d)/d](1 - P_0)$
Average number in the system	L	$L_q + 1 - P_0$
Average number in service	L_s	$1 - P_0$
In-system arrival rate	λ_a	$\lambda(N - L)$
Average time in queue	W_q	L_q/λ_a
Average time is service	W_s	$1/\mu$
Average time in system	W	$L/\lambda_a = W_q + 1/\mu$
Availability	A	$(N - L)/N$

machine is subject to random failures at rate λ when it is running. There is a single server, such as a repair person or operator. When a machine fails it joins the system for repair, waiting in the queue if the server is busy. Service time is exponential at rate μ. If all machines are running the arrival rate is λN. If n machines have failed and have not yet been repaired, the arrival rate is $\lambda(N - n)$ for $n = 0, 1, 2, \ldots, N$. The formulas for this system are shown in Table 38.7.

Application

This model can help us to decide how many machines a single server can support with a desired level of machine availability.

The *M/M/s/G/N/N* System (*s* Servers, Finite Population *N*)

This model has a finite population, N. As with the previous system, a common application of the $M/M/s/G/N/N$ system is where the items are machines and so we will again refer to the items in the population as machines. Each machine is subject to random failures at rate λ when it is running. There are s servers, such as repair persons or operators. When a machine fails it joins the system for repair, waiting in the queue if the servers are busy. Service time is exponential at rate μ. If all machines are running the arrival rate is λN. If n machines have failed and not yet been repaired, the arrival rate is $\lambda(N - n)$ for $n = 0, 1, 2, \ldots, N$. The formulas for this system are shown in Table 38.8.

Table 38.8: $M/M/s/G/N/N$ System Formulas

Description	Symbol	Formula
Traffic density	d	λ/μ
Probability the server is idle	P_0	$1/\left[\sum_{n=0}^{s-1} d^n [N!/(N-n)!n!] + \sum_{n=s}^{N} d^n [N!/(N-n)!s!s^{n-s}]\right]$
Probability n items in the system	P_n	$P_0(d^n[N!/(N-n)!n!])$ for $1 \leq n \leq s$
		$P_0(d^n[N!/(N-n)!s!s^{n-s}])$ for $s < n \leq N$
Average number in queue	L_q	$\sum_{n=s}^{N}(n-s)P_n$
Average number in the system	L	$L_q + \sum_{n=0}^{s-1} nP_n + s\left(1 - \sum_{n=0}^{s-1} P_n\right)$
In-system arrival rate	λ_a	$\lambda(N-L)$
Average time in queue	W_q	L_q/λ_a
Average time is service	W_s	$1/\mu$
Average time in system	W	$L/\lambda_a = W_q + 1/\mu$
Availability	A	$(N-L)/N$

Application

This model can help us to decide how many servers are needed to support a given number of machines at a desired level of machine availability.

Chapter 39

Rayleigh Distribution

Range $0 < x < \infty$.

Scale parameter $b > 0$.

Distribution function	$1 - \exp[-x^2/(2b^2)]$
Probability density function	$(x/b^2)\exp[-x^2/(2b^2)]$
Inverse distribution function (of probability α)	$[-2b^2 \log(1-\alpha)]^{1/2}$
Hazard function	x/b^2
rth Moment about the origin	$(2^{1/2}b)^r(r/2)\Gamma(r/2)$
Mean	$b(\pi/2)^{1/2}$
Variance	$(2-\pi/2)b^2$
Coefficient of skewness	$2(\pi-3)\pi^{1/2}/(4-\pi)^{3/2} \approx 0.63$
Coefficient of kurtosis	$(32-3\pi^2)/(4-\pi)^2 \approx 3.25$
Coefficient of variation	$(4/\pi - 1)^{1/2}$
Mode	b
Median	$b(\log 4)^{1/2}$

The probability density function for the Rayleigh variate $X: b$ is shown in Figure 39.1 for selected values of the scale parameter, b, with the corresponding distributions functions shown in Figure 39.2.

39.1 VARIATE RELATIONSHIPS

1. The Rayleigh variate $X(b)$ corresponds to the Weibull variate $W(b\sqrt{(2)}, 2)$.

2. The square of a Rayleigh variate with parameter $b = 1$ corresponds to the chi-squared variate with 2 degrees of freedom, $\chi : 2$.

3. The square of a Rayleigh variate with parameter b corresponds to an exponential variate with parameter $2b^2$.

Statistical Distributions, Fourth Edition, by Catherine Forbes, Merran Evans, Nicholas Hastings, and Brian Peacock

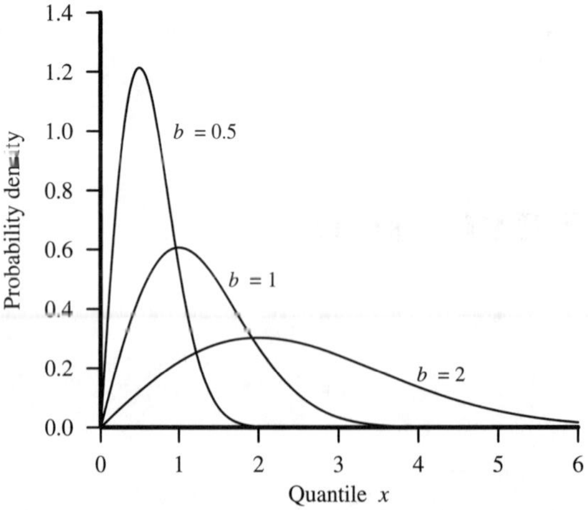

Figure 39.1. Probability density function for the Rayleigh variate; $\chi : b$.

4. The Rayleigh variate with parameter $b = \sigma$, here denoted $\boldsymbol{X}$: σ, is related to independent normal variates $\boldsymbol{N}$: 0, σ by

$$\boldsymbol{X} : \sigma \sim \left[(\boldsymbol{N} : 0, \sigma)_1^2 + (\boldsymbol{N} : 0, \sigma)_2^2\right]^{1/2}.$$

5. A generalization of the Rayleigh variate, related to the sum of squares of ν independent $\boldsymbol{N}$: 0, σ variates, has pdf

$$\frac{2x^{\nu-1}\exp(-x^2/2b^2)}{(2b^2)^{\nu/2}\Gamma(\nu/2)}.$$

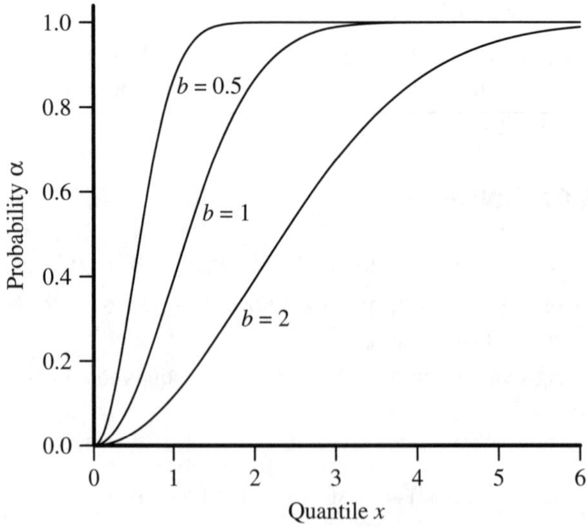

Figure 39.2. Distribution function for the Rayleigh variate; $\chi : b$.

with rth moment about the origin

$$\frac{(2^{1/2}b)^r\Gamma\big((r+v)/2\big)}{\Gamma(v/2)}.$$

For $b = 1$, this corresponds to the chi variate $\chi : v$.

39.2 PARAMETER ESTIMATION

Parameter	Estimator	Method/Properties
b	$\left(\frac{1}{2n}\sum_{i=1}^{n} x_i^2\right)^{1/2}$	Maximum likelihood

Chapter 40

Rectangular (Uniform) Continuous Distribution

Every value in the range of the distribution is equally likely to occur. This is the distribution taken by uniform random numbers. It is widely used as the basis for the generation of random numbers for other statistical distributions.

Variate $\boldsymbol{R}$: $a,\ b$.

Where we write $\boldsymbol{R}$ without specifying parameters, we imply the standard or unit rectangular variate $\boldsymbol{R}$: $0,\ 1$.

Range $a \le x \le b$.

Location parameter a, the lower limit of the range. Parameter b, the upper limit of the range.

Distribution function	$(x-a)/(b-a)$
Probability density function	$1/(b-a)$
Inverse distribution function (of probability α)	$a+\alpha(b-a)$
Inverse survival function (of probability α)	$b-\alpha(b-a)$
Hazard function	$1/(b-x)$
Cumulative hazard function	$-\log[(b-x)/(b-a)]$
Moment generating function	$[\exp(bt)-\exp(at)]/[t(b-a)]$
Characteristic function	$[\exp(ibt)-\exp(iat)]/[it(b-a)]$
rth Moment about the origin	$\dfrac{b^{r+1}-a^{r+1}}{(b-a)(r+1)}$
Mean	$(a+b)/2$

Statistical Distributions, Fourth Edition, by Catherine Forbes, Merran Evans, Nicholas Hastings, and Brian Peacock

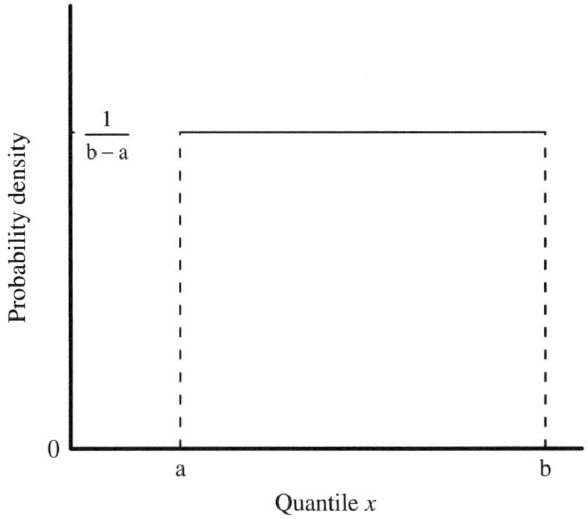

Figure 40.1. Probability density function for the rectangular variate $\boldsymbol{R}$: a, b.

rth Moment about the mean	$\begin{cases} 0, & r \text{ rod} \\ \dfrac{[(b-a)/2]^r}{(r+1)}, & r \text{ even} \end{cases}$
Variance	$(b-a)^2/12$
Mean deviation	$(b-a)/4$
Median	$(a+b)/2$
Coefficient of skewness	0
Coefficient of kurtosis	9/5
Coefficient of variation	$(b-a)/\left[(b+a)3^{1/2}\right]$
Information content	$\log_2 b$

The probability density function for the $\boldsymbol{R}$: a, b variate is shown in Figure 40.1, with the corresponding distribution function shown in Figures 40.2. The hazard function for the $\boldsymbol{R}$: 0, 1 variate is shown in Figure 40.3.

40.1 VARIATE RELATIONSHIPS

1. Let X be any variate and G_X be the inverse distribution function of X, that is,

$$\Pr\left[X \leq G_X(\alpha)\right] = \alpha, \qquad 0 \leq \alpha \leq 1.$$

Variate X is related to the unit rectangular variate $\boldsymbol{R}$ by

$$X \sim G_X(\boldsymbol{R}).$$

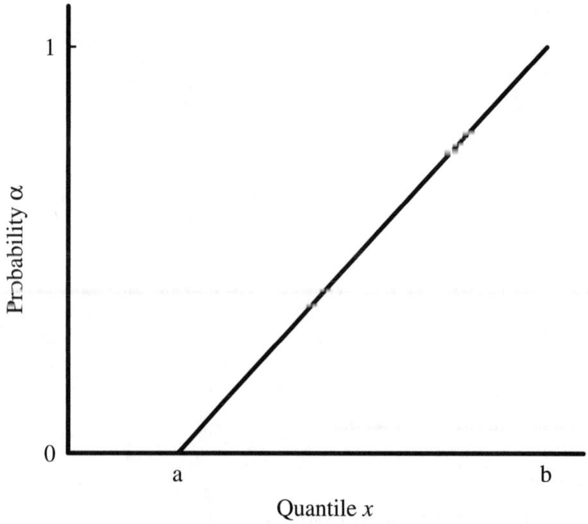

Figure 40.2. Distribution function for the rectangular variate $\boldsymbol{R}$: a, b.

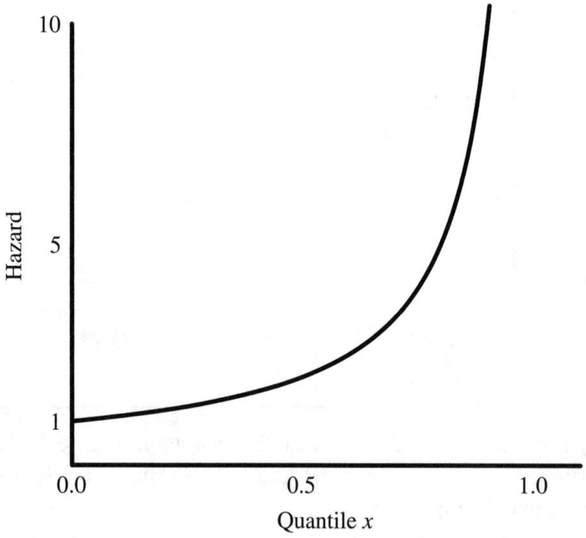

Figure 40.3. Hazard function for the unit rectangular variate $\boldsymbol{R}$: 0, 1.

For $\boldsymbol{X}$ any variate with a continuous density function $f_X(x)$, the cdf of x, $f_X(x)$, satisfies

$$F_X(x) \sim \boldsymbol{R} : 0, 1.$$

2. The distribution function of the sum of n-independent unit rectangular variates $\boldsymbol{R}_i, i = 1, \ldots, n$ is

$$\sum_{i=0}^{k} (-1)^i \binom{n}{i} \frac{(x-i)^n}{n!}, \quad k \leq x \leq k+1, \quad k = 0, 1, \ldots, n-1.$$

3. The unit parameter beta variate $\boldsymbol{\beta}$: 1, 1 and the power function variate, here denoted $\boldsymbol{X}$: 1, 1, correspond to a unit rectangular variate $\boldsymbol{R}$.

4. The mean of two independent unit rectangular variates is a standard symmetrical triangular variate.

40.2 PARAMETER ESTIMATION

Parameter		Estimator	Method
Lower limit,	a	$\bar{x} - 3^{1/2}s$	Matching moments
Upper limit,	b	$\bar{x} + 3^{1/2}s$	Matching moments

40.3 RANDOM NUMBER GENERATION

Algorithms to generate pseudorandom numbers, which closely approximate independent standard unit rectangular variates, $\boldsymbol{R}$: 0, 1, are a standard feature in statistical software.

Chapter 41

Rectangular (Uniform) Discrete Distribution

In sample surveys it is often assumed that the items (e.g., people) being surveyed are uniformly distributed through the sampling frame.

Variate $\boldsymbol{D}$: $0, n$.

Range $0 \leq x \leq n$, x an integer taking values $0, 1, 2, \ldots, n$.

Distribution function	$(x+1)/(n+1)$
Probability function	$1/(n+1)$
Inverse distribution function (of probability α)	$\alpha(n+1)-1$
Survival function	$(n-x)/(n+1)$
Inverse survival function (of probability α)	$n-\alpha(n+1)$
Hazard function	$1/(n-x)$
Probability generating function	$(1-t^{n+1})/[(n+1)(1-t)]$
Characteristic function	$\{1-\exp[it(n+1)]\}/$ $\{[1-\exp(it)](n+1)\}$
Moments about the origin	
Mean	$n/2$
Second	$n(2n+1)/6$
Third	$n^2(n+1)/4$
Variance	$n(n+2)/12$
Coefficient of skewness	0
Coefficient of kurtosis	$\frac{3}{5}[3-4/n(n+2)]$
Coefficient of variation	$[(n+2)/3n]^{1/2}$

The probability function for the $\boldsymbol{D}: 0, n$ variate is shown in Figure 41.1, with the corresponding distribution function shown in Figure 41.2.

Statistical Distributions, Fourth Edition, by Catherine Forbes, Merran Evans, Nicholas Hastings, and Brian Peacock

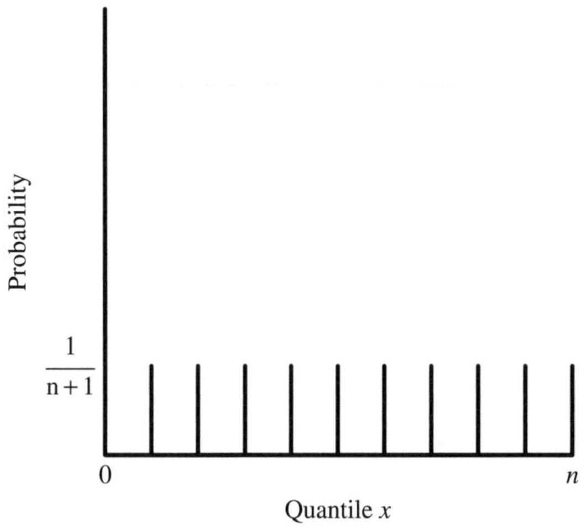

Figure 41.1. Probability function for the discrete rectangular variate $\boldsymbol{D}$: 0, n.

41.1 GENERAL FORM

Let $a \leq x \leq a + nh$, such that any point of the sample space is equally likely. The term a is a location parameter and h, the size of the increments, is a scale parameter. The probability function is still $1/(n+1)$. The mean is $a + nh/2$, and the rth moments are those of the standard form $\boldsymbol{D}$: 0, 1 multiplied by h^r.

As N tends to infinity and h tends to zero with $nh = b - a$, the discrete rectangular variate $\boldsymbol{D}$: a, $a + nh$ tends to the continuous rectangular variate $\boldsymbol{R}$: a, b.

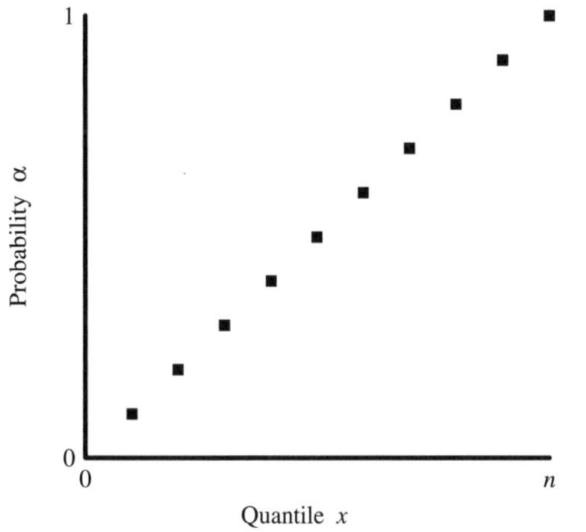

Figure 41.2. Distribution function for the discrete rectangular variate $\boldsymbol{D}$: 0, n.

41.2 PARAMETER ESTIMATION

Parameter	Estimator	Method/Properties
Location parameter, a	$\bar{x} - nh/2$	Matching moments
Increments, h	$\left\{12s^2/[n(n+2)]\right\}^{1/2}$	Matching moments

Chapter 42

Student's *t* Distribution

The Student's t distribution is used to test whether the difference between the means of two samples of observations is statistically significant. For example, the heights of a random sample of basketball players could be compared with the heights from a random sample of football players. The Student's t distribution would be used to test whether the data indicated that one group was significantly taller than the other. More precisely, it would be testing the hypothesis that both samples were drawn from the same normal population. A significant value of t would cause the hypothesis to be rejected, indicating that the means were significantly different.

Variate $t: \nu$.

Range $-\infty < x < \infty$.

Shape parameter ν, degrees of freedom, ν a positive integer.

Distribution function

$$\begin{cases} \dfrac{1}{2} + \dfrac{1}{\pi}\tan^{-1}\left(\dfrac{x}{\nu^{1/2}}\right) \\ \quad + \dfrac{1}{\pi}\dfrac{x\nu^{1/2}}{\nu + x^2}\left[\displaystyle\sum_{j=0}^{(\nu-3)/2} \dfrac{a_j}{(1 + x^2/\nu)^j}\right], & \nu \text{ odd} \\ \dfrac{1}{2} + \dfrac{x}{2(\nu + x^2)^{1/2}} \displaystyle\sum_{j=0}^{(\nu-2)/2} \dfrac{b_j}{\left(1 + \dfrac{x^2}{\nu}\right)^j}, & \nu \text{ even} \\ \text{where } a_j = [2j/(2j+1)]a_{j-1}, \quad a_0 = 1 \\ \qquad\quad b_j = [(2j-1)/2j]b_{j-1}, \quad b_0 = 1 \end{cases}$$

Probability density function

$$\frac{\{\Gamma[(\nu+1)/2]\}}{(\pi\nu)^{1/2}\Gamma(\nu/2)[1 + (x^2/\nu)]^{(\nu+1)/2}}$$

Mean $\quad 0, \quad \nu > 1$

Statistical Distributions, Fourth Edition, by Catherine Forbes, Merran Evans, Nicholas Hastings, and Brian Peacock

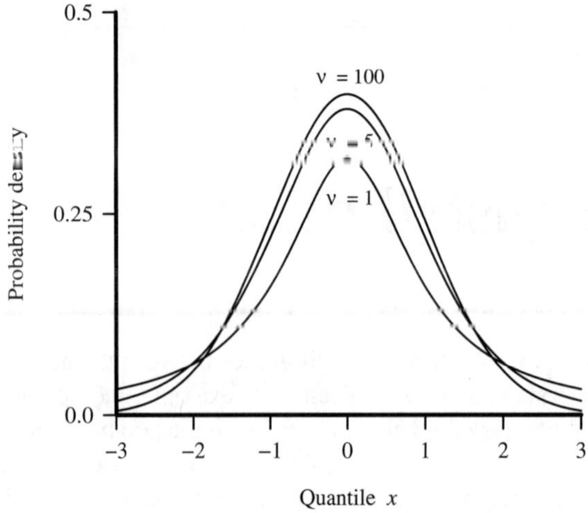

Figure 42.1. Probability density function for Student's t variate, $\boldsymbol{t}$: ν.

rth Moment about the mean
$$\begin{cases} \mu_r = 0, & r \text{ odd} \\ \mu_r = \dfrac{1 \cdot 3 \cdot 5 \cdots (r-1)\nu^{r/2}}{(\nu-2)(\nu-4)\cdots(\nu-r)} & r \text{ even}, \quad \nu > r \end{cases}$$

Variance $\quad \nu/(\nu-2), \quad \nu > 2$

Mean deviation $\quad \nu^{1/2}\Gamma\left[\frac{1}{2}(\nu-1)\right] \Big/ \left[\pi^{1/2}\Gamma\left(\frac{1}{2}\nu\right)\right]$

Mode $\quad 0$

Coefficient of skewness $\quad 0, \quad \nu > 3 \quad$ (always symmetrical)

Coefficient of kurtosis $\quad 3(\nu-2)/(\nu-4), \quad \nu > 4$

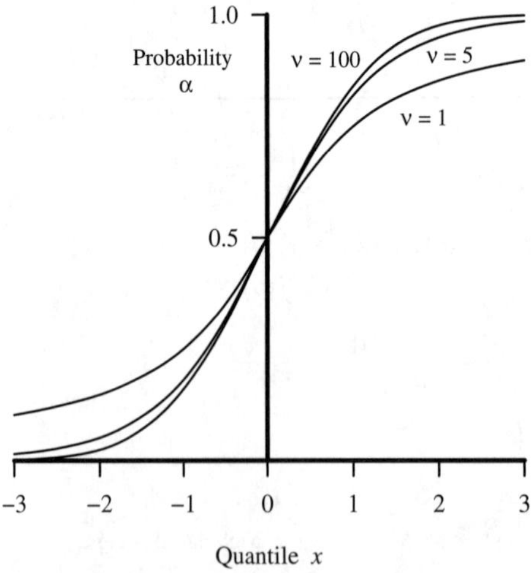

Figure 42.2. Distribution function for Student's t variate, $\boldsymbol{t}$: ν.

The probability density function for the Student's t variate $\boldsymbol{t}: \nu$ is shown in Figure 42.1 for selected values of the degrees of freedom parameter, ν, with the corresponding distributions functions shown in Figure 42.2.

42.1 VARIATE RELATIONSHIPS

1. The Student's t variate with ν degrees of freedom, $\boldsymbol{t}: \nu$, is related to the independent chi-squared variate $\boldsymbol{\chi}^2 : \nu$, the F variate $\boldsymbol{F}: 1, \nu$, and the unit normal variate $\boldsymbol{N}: 0, 1$ by

$$\begin{aligned}(\boldsymbol{t}:\nu)^2 &\sim (\boldsymbol{\chi}^2 : 1)/\left[(\boldsymbol{\chi}^2 : \nu)/\nu\right]\\ &\sim \boldsymbol{F} : 1, \nu\\ &\sim (\boldsymbol{N} : 0, 1)^2/\left[(\boldsymbol{\chi}^2 : \nu)/\nu\right]\\ \boldsymbol{t} : \nu &\sim (\boldsymbol{N} : 0, 1)/\left[(\boldsymbol{\chi}^2 : \nu)/\nu\right]^{1/2}.\end{aligned}$$

Equivalently, in terms of a probability statement,

$$\Pr[(\boldsymbol{t} : \nu) \leq x] = \frac{1}{2}\left\{1 + \Pr\left[(\boldsymbol{F} : 1, \nu) \leq x^2\right]\right\}.$$

In terms of the inverse survival function of $\boldsymbol{t} : \nu$ at probability level $\frac{1}{2}\alpha$, denoted $Z_t\left(\frac{1}{2}\alpha : \nu\right)$, and the survival function of the $\boldsymbol{F}$ variate $\boldsymbol{F} : 1, \nu$ at probability level α, denoted $Z_{\boldsymbol{F}}(\alpha : 1, \nu)$, the last equation is equivalent to

$$\left[Z_t\left(\frac{1}{2}\alpha : \nu\right)\right]^2 = Z_{\boldsymbol{F}}(\alpha : 1, \nu).$$

2. As ν tends to infinity, the variate $\boldsymbol{t} : \nu$ tends to the unit normal variate $\boldsymbol{N}$: 0,1. The approximation is reasonable for $\nu \geq 30$.

$$\boldsymbol{t} : \nu \approx \boldsymbol{N} : 0, 1; \quad \nu \geq 30.$$

3. Consider independent normal variates $\boldsymbol{N} : \mu, \sigma$. Define variates $\bar{x}, s^2$ as follows:

$$\bar{x} \sim \left(\frac{1}{n}\right)\sum_{i=1}^{n}(\boldsymbol{N} : \mu, \sigma)_i, \quad s^2 \sim \left(\frac{1}{n}\right)\sum_{i=1}^{n}[(\boldsymbol{N} : \mu, \sigma)_i - \bar{x}]^2.$$

Then

$$\boldsymbol{t} : n - 1 \sim \frac{\bar{x} - \mu}{s/(n-1)^{1/2}}.$$

4. Consider a set of n_1-independent normal variates $\boldsymbol{N} : \mu_1, \sigma$, and a set of n_2-independent normal variates $\boldsymbol{N} : \mu_2, \sigma$. Define variates $\bar{x}_1, \bar{x}_2, s_1^2, s_2^2$ as follows:

$$\bar{x}_1 \sim \left(\frac{1}{n_1}\right)\sum_{i=1}^{n_1}(\boldsymbol{N}:\mu_1,\sigma)_i,$$

$$\bar{x}_2 \sim \left(\frac{1}{n_2}\right)\sum_{j=1}^{n_2}(\boldsymbol{N}:\mu_2,\sigma)_j$$

$$s_1^2 \sim \left(\frac{1}{n_1}\right)\sum_{i=1}^{n_1}\left[(\boldsymbol{N}:\mu_1,\sigma)_i-\bar{x}_1\right]^2$$

$$s_2^2 \sim \left(\frac{1}{n_2}\right)\sum_{j=1}^{n_2}\left[(\boldsymbol{N}:\mu_2,\sigma)_j-\bar{x}_2\right]^2.$$

Then

$$\boldsymbol{t}:n_1+n_2-2 \sim \frac{(\bar{x}_1-\bar{x}_2)-(\mu_1-\mu_2)}{\left(\dfrac{n_1s_1^2+n_2s_2^2}{n_1+n_2-2}\right)^{1/2}\left(\dfrac{1}{n_1}+\dfrac{1}{n_2}\right)^{1/2}}.$$

5. The $\boldsymbol{t}$: 1 variate corresponds to the standard Cauchy variate, $C:0,1$.

6. The $\boldsymbol{t}:\nu$ variate is related to two independent $\boldsymbol{F}:\nu,\nu$ variates by

$$(\nu^{1/2}/2)\left[(\boldsymbol{F}:\nu,\nu)_1^{1/2}-(\boldsymbol{F}:\nu,\nu)_2^{-1/2}\right]\sim \boldsymbol{t}:\nu.$$

7. Two independent chi-squared variates $\chi^2:\nu$ are related to the $\boldsymbol{t}:\nu$ variate by

$$\left(\frac{\nu^{1/2}}{2}\right)\frac{\left[(\chi^2:\nu)_1-(\chi^2:\nu)_2\right]}{\left[(\chi^2:\nu)_1(\chi^2:\nu)_2\right]^{1/2}}\sim \boldsymbol{t}:\nu.$$

42.2 RANDOM NUMBER GENERATION

From independent $\boldsymbol{N}$: 0, 1 and $\chi^2:\nu$ variates

$$\boldsymbol{t}:\nu \sim \frac{\boldsymbol{N}:0,1}{\sqrt{(\chi^2:\nu)/\nu}}$$

or from a set of $\nu+1$ independent $\boldsymbol{N}$: 0, 1 variates

$$\boldsymbol{t}:\nu \sim \frac{(\boldsymbol{N}:0,1)_{\nu+1}}{\sqrt{\sum_{i=1}^{\nu}(\boldsymbol{N}:0,1)_i^2/\nu}}.$$

Chapter 43

Student's t (Noncentral) Distribution

The noncentral t distribution can be applied to the testing of one-sided hypotheses related to normally distributed data. For example, if the mean test score of a large cohort of students is known, then it would be possible to test the hypothesis that the mean score of a subset, from a particular teacher, was higher than the general mean by a specified amount.

Variate t: ν, δ.

Range $-\infty < x < \infty$.

Shape parameters ν a positive integer, the degrees of freedom and $-\infty < \delta < \infty$, the noncentrality parameter.

Probability density function
$$\frac{(\nu)^{\nu/2}\exp(-\delta^2/2)}{\Gamma(\nu/2)\pi^{1/2}(\nu+x^2)^{(\nu+1)/2}} \times \sum_{i=0}^{\infty}\Gamma\left(\frac{\nu+i+1}{2}\right)\frac{(x\delta)^i}{i!}\left(\frac{2}{\nu+x^2}\right)^{i/2}$$

rth Moment about the origin
$$(\nu/2)^{r/2}\Gamma((\nu-r)/2)/\Gamma(\nu/2) \times \sum_{j=0}^{r/2}\binom{r}{2j}(2j)!\delta^{r-2j}/(2^j j!),\quad \nu > r$$

Mean
$$\frac{\delta(\nu/2)^{1/2}\Gamma((\nu-1)/2)}{\Gamma(\nu/2)},\quad \nu > 1$$

Variance
$$\frac{\nu}{(\nu-2)}(1+\delta^2) - \frac{\nu}{2}\delta^2\left(\frac{\Gamma((\nu-1)/2)}{\Gamma(\nu/2)}\right)^2,\quad \nu > 2$$

Statistical Distributions, Fourth Edition, by Catherine Forbes, Merran Evans, Nicholas Hastings, and Brian Peacock

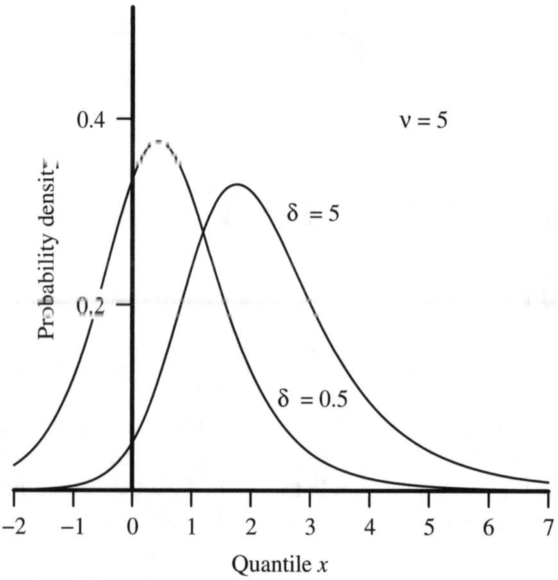

Figure 43.1. Probability density function for the (noncentral) Student's t variate $\boldsymbol{t}$: ν, δ.

The probability density function for the noncentral Student's t variate $\boldsymbol{t} : 5, \delta$ is shown in Figure 43.1 for selected values of the noncentrality parameter, δ.

43.1 VARIATE RELATIONSHIPS

1. The noncentral t variate $\boldsymbol{t}$: ν, δ is related to the independent chi-squared variate, $\chi^2 : \nu$, and normal variate, $\boldsymbol{N}$: 0, 1 (or $\boldsymbol{N}$: δ, 1), by

$$\boldsymbol{t} : \nu, \delta \sim \frac{(\boldsymbol{N} : 0, 1) + \delta}{[(\boldsymbol{\chi}^2 : \nu)/\nu]^{1/2}} \sim \frac{\boldsymbol{N} : \delta, 1}{[(\boldsymbol{\chi}^2 : \nu)/\nu]^{1/2}}.$$

2. The noncentral t variate $\boldsymbol{t}$: ν, δ is the same as the (central) Student's $\boldsymbol{t}$ variate $\boldsymbol{t}$: ν for $\delta = 0$.

3. The noncentral t variate $\boldsymbol{t}$: ν, δ is related to the noncentral beta variate $\boldsymbol{\beta}$: 1, ν, δ^2 with parameters 1, ν, and δ by

$$\boldsymbol{\beta} : 1, \nu, \delta^2 \sim (\boldsymbol{t} : \nu, \delta)^2 / \left[\nu + (\boldsymbol{t} : \nu, \delta)^2\right].$$

Chapter 44

Triangular Distribution

Range $a \leq x \leq b$.

Parameters: Shape parameter c, the mode.

Location parameter a, the lower limit; parameter b, the upper limit.

Distribution function	$\begin{cases} \dfrac{(x-a)^2}{(b-a)(c-a)}, & \text{if } a \leq x \leq c \\ 1 - \dfrac{(b-x)^2}{(b-a)(b-c)}, & \text{if } c \leq x \leq b \end{cases}$
Probability density function	$\begin{cases} 2(x-a)/[(b-a)(c-a)], & \text{if } a \leq x \leq c \\ 2(b-x)/[(b-a)(b-c)], & \text{if } c \leq x \leq b \end{cases}$
Mean	$(a+b+c)/3$
Variance	$(a^2+b^2+c^2-ab-ac-bc)/18$
Mode	c

The probability density function for the triangular variate for selected values of the shape parameter, c, relative to the two location parameters, a and b, are shown in Figure 44.1.

44.1 VARIATE RELATIONSHIPS

1. The standard triangular variate corresponding to $a = 0, b = 1$, has median $\sqrt{c/2}$ for $c \geq \frac{1}{2}$ and $1 - \sqrt{(1-c)/2}$ for $c \leq \frac{1}{2}$.

2. The standard symmetrical triangular variate is a special case of the triangular variate with $a = 0, b = 1, c = \frac{1}{2}$. It has even moments about the mean

Statistical Distributions, Fourth Edition, by Catherine Forbes, Merran Evans, Nicholas Hastings, and Brian Peacock

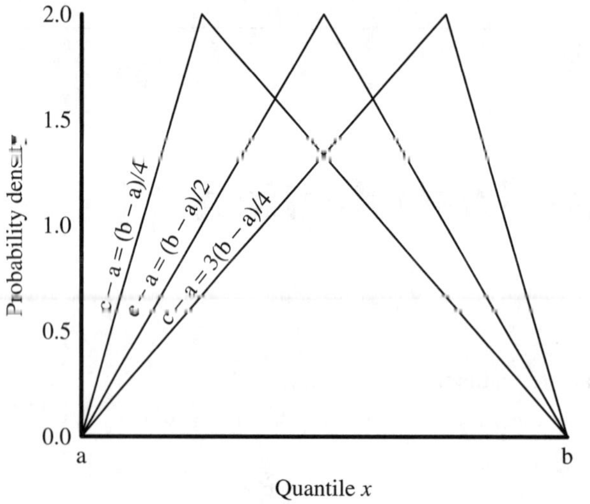

Figure 44.1. Probability density function for the triangular variate.

$\mu_r = [2^{r-1}(r+1)(r+2)]^{-1}$ and odd moments zero. The skewness coefficient is zero and kurtosis 12/5.

44.2 RANDOM NUMBER GENERATION

The standard symmetrical triangular variate is generated from independent unit rectangular variates $\boldsymbol{R}_1, \boldsymbol{R}_2$ by

$$(\boldsymbol{R}_1 + \boldsymbol{R}_2)/2.$$

Chapter 45

von Mises Distribution

Range $0 < x \leq 2\pi$, where x is a circular random variate.

Scale parameter $b > 0$ is the concentration parameter.

Location parameter $0 \leq a < 2\pi$ is the mean direction.

Distribution function	$[2\pi I_0(b)]^{-1}\left\{xI_0(b) + 2\sum_{j=0}^{\infty} I_j(b)\frac{\sin j(x-a)}{j}\right\}$,
	where $I_t(b) = \left(\frac{b}{2}\right)^t \sum_{i=0}^{\infty} \frac{(b^2/4)^t}{i!\Gamma(t+i+1)}$
	is the modified Bessel function of the first kind of order t;
	for order $t = 0$, $I_0(b) = \sum_{i=0}^{\infty} b^{2i}/[2^{2i}(i!)^2]$
Probability density function	$\exp[b\cos(x-a)]/(2\pi I_0(b))$
Characteristic function	$[I_t(b)/I_0][\cos(at) + i\sin(at)]$
rth Trigonometric moment about the origin	$\begin{cases}[I_r(b)/I_0(b)]\cos(ar)\\ [I_r(b)/I_0(b)]\sin(ar)\end{cases}$
Mean direction	a
Mode	a
Circular variance	$1 - I_1(b)/I_0(b)$

45.1 NOTE

The von Mises distribution can be regarded as the circular analogue of the normal distribution on the line. The distribution is unimodal, symmetric about a, and infinitely divisible. The minimum value occurs at $a \pm \pi$ [whichever is in range $(0, 2\pi)$], and the ratio of maximum to minimum values of the pdf is $\exp(2b)$.

Statistical Distributions, Fourth Edition, by Catherine Forbes, Merran Evans, Nicholas Hastings, and Brian Peacock

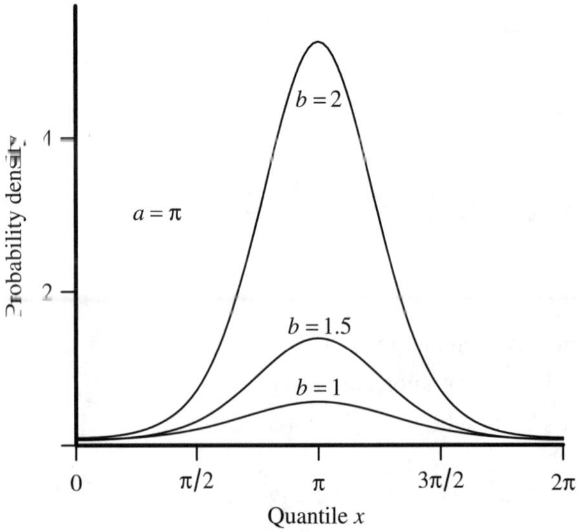

Figure 45.1. Probability density function for the von Mises variate.

The probability density function for the von Mises variate with location parameter $a = \pi$ and for selected values of the scale parameter, b, are shown in Figure 45.1.

45.2 VARIATE RELATIONSHIPS

1. For $b = 0$, the von Mises variate reduces to the rectangular variate ***R***: *a, b* with $a = 0, b = 2\pi$ with pdf $1/(2\pi)$.

2. For large b, the von Mises variate tends to the normal variate ***N***: μ, σ with $\mu = 0, \sigma = 1/b$.

3. For independent normal variates, with means $\sin(a)$ and $\cos(a)$, respectively, let their corresponding polar coordinates be R and θ. The conditional distribution of θ, given $R = 1$, is the von Mises distribution with parameters a and b.

45.3 PARAMETER ESTIMATION

Parameter	Estimator	Method/Properties
a	$\tan^{-1}\left(\sum_{i=1}^{n}\sin x_i \Big/ \sum_{i=1}^{n}\cos x_i\right)$	Maximum likelihood
$I_1(b)/I_0(b)$ (a measure of precision)	$\frac{1}{n}\left[\left(\sum_{i=1}^{n}\cos x_i\right)^2 + \left(\sum_{i=1}^{n}\sin x_i\right)^2\right]^{1/2}$	Maximum likelihood

Chapter 46

Weibull Distribution

The Weibull variate is commonly used as a lifetime distribution in reliability applications. The two-parameter Weibull distribution can represent decreasing, constant, or increasing failure rates. These correspond to the three sections of the "bathtub curve" of reliability, referred to also as "burn-in," "random," and "wearout" phases of life. The bi-Weibull distribution can represent combinations of two such phases of life.

Variate $\boldsymbol{W} : \eta, \beta$.

Range $0 \le x < \infty$.

Scale parameter $\eta > 0$ is the characteristic life.

Shape parameter $\beta > 0$.

Distribution function	$1 - \exp[-(x/\eta)^{\beta}]$
Probability density function	$(\beta x^{\beta-1}/\eta^{\beta})\exp[-(x/\eta)^{\beta}]$
Inverse distribution function (of probability α)	$\eta\{\log[1/(1-\alpha)]\}^{1/\beta}$
Survival function	$\exp[-(x/\eta)^{\beta}]$
Inverse survival function (of probability α)	$\eta[\log(1/\alpha)^{1/\beta}]$
Hazard function	$\beta x^{\beta-1}/\eta^{\beta}$
Cumulative hazard function	$(x/\eta)^{\beta}$
rth Moment about the origin	$\eta^{r}\Gamma[(\beta+r)/\beta]$
Mean	$\eta\Gamma[(\beta+1)/\beta]$
Variance	$\eta^{2}(\Gamma[(\beta+2)/\beta] - \{\Gamma[(\beta+1)/\beta]\}^{2})$
Mode	$\begin{cases} \eta(1-1/\beta)^{1/\beta}, & \beta \ge 1 \\ 0, & \beta \le 1 \end{cases}$

Statistical Distributions, Fourth Edition, by Catherine Forbes, Merran Evans, Nicholas Hastings, and Brian Peacock

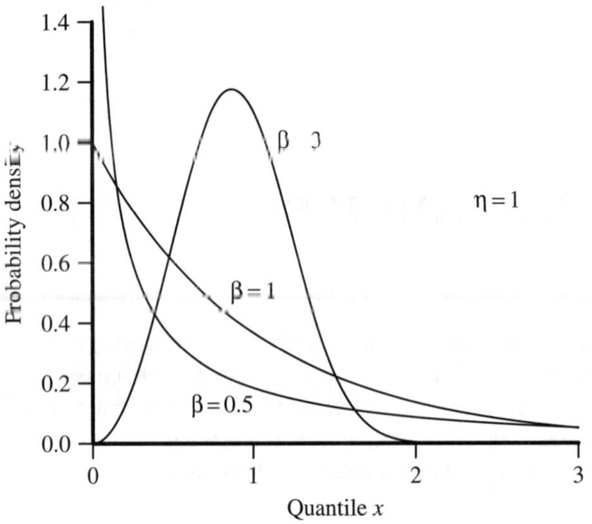

Figure 46.1. Probability density function for the Weibull variate. $\boldsymbol{W}: \eta,\ \beta$.

Median	$\eta(\log 2)^{1/\beta}$
Coefficient of variation	$\left(\dfrac{\Gamma[(\beta+2)/\beta]}{\{\Gamma[(\beta+1)/\beta]\}^2} - 1\right)^{1/2}$

The probability density function for the $\boldsymbol{W}$: 1, β variate for selected values of the shape parameter, β, is shown in Figure 46.1, with the corresponding distribution functions shown in Figure 46.2 and the corresponding hazard functions shown in Figure 46.3. The relationship between the shape parameter, β, and the mean and standard deviation of the $\boldsymbol{W}$: 1, β variate is shown in Figure 46.4.

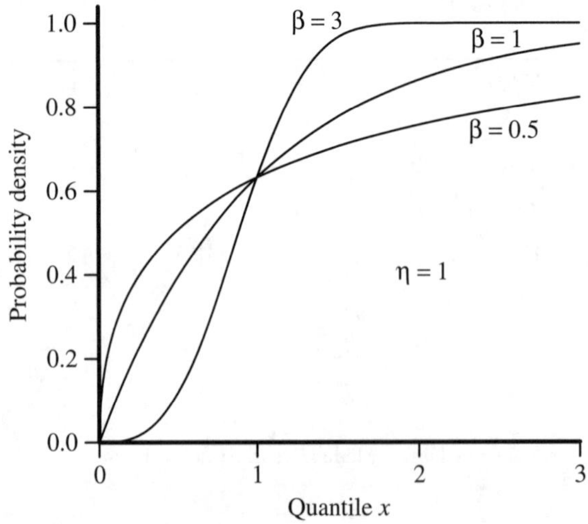

Figure 46.2. Distribution function for the Weibull variate. $\boldsymbol{W}: \eta,\ \beta$.

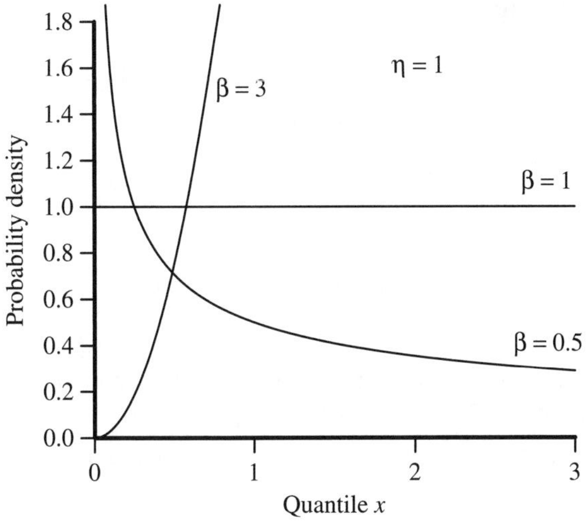

Figure 46.3. Hazard function for the Weibull variate $W : \eta,\ \beta$.

46.1 NOTE

The characteristic life η has the property that

$$\Pr[(W : \eta, \beta) \leq \eta] = 1 - \exp(-1) = 0.632 \text{ for every } \beta.$$

Thus η is approximately the 63rd percentile.

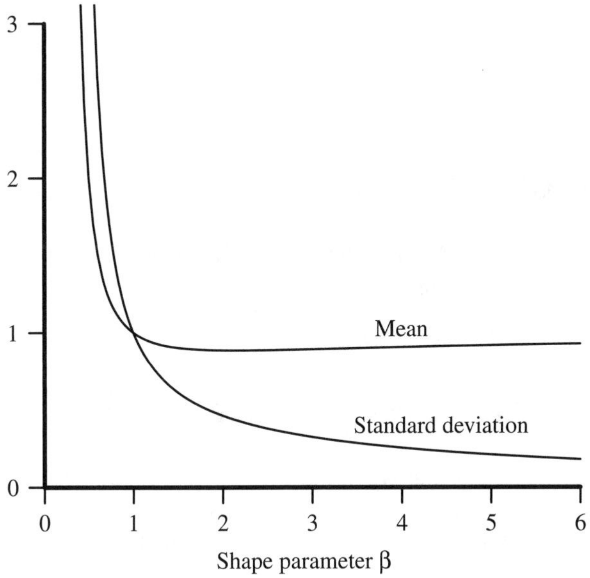

Figure 46.4. Weibull $W : 1, \beta$ mean and standard deviation as a function of the shape parameter β.

46.2 VARIATE RELATIONSHIPS

1. The standard Weibull variate has a characteristic life value $\eta = 1$. The general Weibull variate $\boldsymbol{W}$: η, β is a sealed standard Weibull variate.

$$\boldsymbol{W} : \eta, \beta \sim \eta(\boldsymbol{W} : 1, \beta),$$

2. The Weibull variate $\boldsymbol{W} : \eta, \beta$ with shape parameter $\beta = 1$ is the exponential variate $\boldsymbol{E} : \eta$ with mean η,

$$\boldsymbol{W} : \eta, 1 \sim \boldsymbol{E} : \eta.$$

The Weibull variate $\boldsymbol{W} : \eta, \beta$ is related to $\boldsymbol{E} : \eta$ by $(\boldsymbol{W} : \eta, \beta)^{\beta} \sim \boldsymbol{E} : \eta^{\beta}$.

3. The Weibull variate $\boldsymbol{W} : \eta, 2$ is the Rayleigh variate, and the Weibull variate $\boldsymbol{W} : \eta, \beta$ is also known as the truncated Rayleigh variate.

4. The Weibull variate $\boldsymbol{W} : \eta, \beta^{\eta}$ is related to the standard extreme value (Gumbel) variate $\boldsymbol{V}$: 0, 1 by

$$-\beta \log[(\boldsymbol{W} : \eta, \beta)/\eta] \sim \boldsymbol{V} : 0, 1.$$

46.3 PARAMETER ESTIMATION

By the method of maximum likelihood the estimators, $\hat{\eta}$, $\hat{\beta}$, of the shape and scale parameters are the solution of the simultaneous equations:

$$\hat{\eta} = \left[\left(\frac{1}{n}\right) \sum_{i=1}^{n} x_i^{\hat{\beta}}\right]^{1/\hat{\beta}}$$

$$\hat{\beta} = \frac{n}{(1/\hat{\eta}) \sum_{i=1}^{n} x_i^{\hat{\beta}} \log x_i - \sum_{i=1}^{n} \log x_i}.$$

46.4 RANDOM NUMBER GENERATION

Random numbers of the Weibull variate $\boldsymbol{W} : \eta, \beta$ can be generated from those of the unit rectangular variate $\boldsymbol{R}$ using the relationship

$$\boldsymbol{W} : \eta, \beta \sim \eta(-\log \boldsymbol{R})^{1/\beta}.$$

46.5 THREE-PARAMETER WEIBULL DISTRIBUTION

Further flexibility can be introduced into the Weibull distribution by adding a third parameter, which is a location parameter and is usually denoted by the symbol gamma (γ). The probability density is zero for $x < \gamma$ and then follows a Weibull distribution with origin at γ. In reliability applications, gamma is often referred to as the minimum life, but this does not guarantee that no failures will occur below this value in the future.

Variate $\boldsymbol{W}: \gamma, \eta, \beta$.

Location parameter $\gamma > 0$.

Scale parameter $\eta > 0$.

Shape parameter $\beta > 0$.

Range $\gamma \leq x \leq +\infty$.

Cumulative distribution function	$1 - \exp\{-[(x-\gamma)/\eta]^{\beta}\}, \quad x \geq \gamma$
Probability density function	$[\beta(x-\gamma)^{\beta-1}/\eta^{\beta}]\exp\{-[(x-\gamma)/\eta]^{\beta}\}, \quad x \geq \gamma$
Inverse distribution function (of probability α)	$\gamma + \eta\{\log[1/(1-\alpha)]\}^{1/\beta}$
Survival function	$\exp\{-[(x-\gamma)/\eta]^{\beta}\}, \quad x \geq \gamma$
Inverse survival function (of probability α)	$\gamma + \eta[\log(1/\alpha)]^{1/\beta}$
Hazard function (failure rate)	$\beta(x-\gamma)^{\beta-1}/\eta^{\beta}, \quad x \geq \gamma$
Cumulative hazard function	$[(x-\gamma)/\eta]^{\beta}, \quad x \geq \gamma$
Mean	$\gamma + \eta\Gamma[(\beta+1)/\beta]$
Variance	$\eta^2(\Gamma[(\beta+2)/\beta] - \{\Gamma[(\beta+1)/\beta]\}^2)$
Mode	$\begin{cases} \gamma + \eta(1-1/\beta)^{1/\beta}, & \beta \geq 1 \\ \gamma, & \beta \leq 1 \end{cases}$

The probability density function for the three parameter $\boldsymbol{W}$: 1, 2, 2 variate is shown in Figure 46.5, with the corresponding distribution function shown in Figure 46.6.

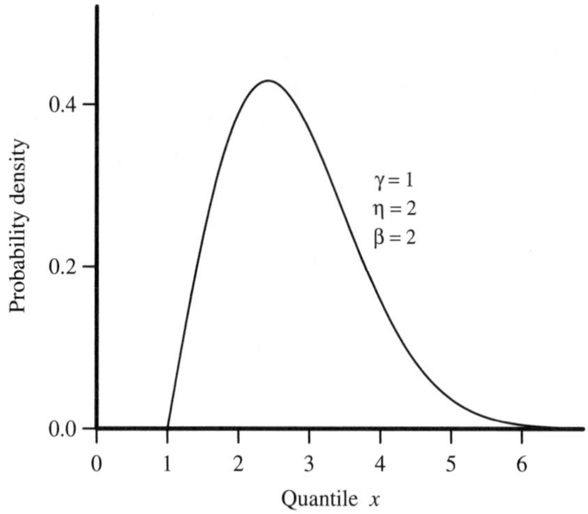

Figure 46.5. Probability density function for the three parameter Weibull variate $\boldsymbol{W}: \gamma, \eta, \beta$.

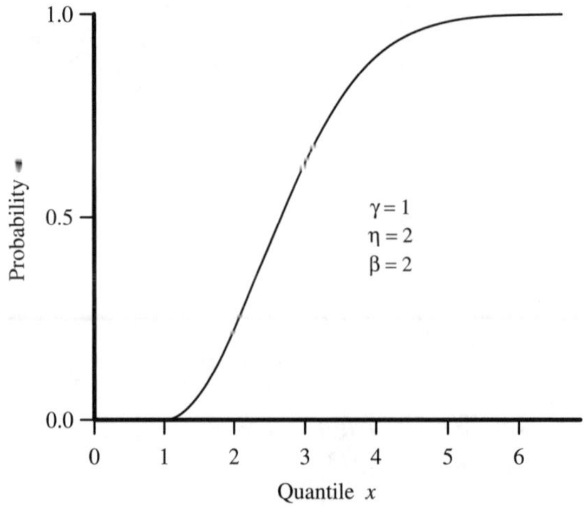

Figure 46.6. Distribution function for the three parameter Weibull variate $\boldsymbol{W}$: γ, η, β.

46.6 THREE-PARAMETER WEIBULL RANDOM NUMBER GENERATION

Random numbers of the Weibull variate $\boldsymbol{W}$: γ, η, β can be generated from the unit rectangular variate $\boldsymbol{R}$ using the relationship

$$\boldsymbol{W} : \gamma, \eta, \beta \sim \gamma + \eta(-\log \boldsymbol{R})^{1/\beta}.$$

46.7 BI-WEIBULL DISTRIBUTION

A bi-Weibull distribution is formed by combining two Weibull distributions. This provides a distribution model with a very flexible shape. In reliability analysis, for example, it can represent combinations of (two of) decreasing, constant, and/or increasing failure rates. Even more flexibility can be introduced by adding more than two Weibull distributions, but this increases the number of parameters to be estimated. Several versions of the bi-Weibull distribution have been proposed by different authors. These versions differ in the way in which the two Weibull distributions are combined, and in the number of parameters specified. The five-parameter bi-Weibull distribution described here is used in the RELCODE software package (www.albanyint.com.au).

46.8 FIVE-PARAMETER BI-WEIBULL DISTRIBUTION

$$\boldsymbol{W} : \lambda, \theta, \gamma, \eta, \beta.$$

Phase 1 scale parameter $\lambda > 0$, shape parameter $\theta > 0$.

Phase 2 location parameter $\gamma \geq 0$, scale parameter $\eta > 0$, shape parameter $\beta > 0$.

A five-parameter bi-Weibull distribution is derived by adding two Weibull *hazard* functions. The first of these hazard functions is a two-parameter Weibull hazard function with the equation

$$h(x) = \lambda\theta(\lambda x)^{\theta-1}$$

where x is the component age, $h(x)$ is the hazard function at age x, λ is the reciprocal of a scale parameter, and θ is a shape parameter. The case where $\theta = 1$ corresponds to a constant failure rate λ. The second hazard function is a three-parameter Weibull hazard function, which becomes operative for $x > \gamma$. The equation is

$$h(x) = \left(\frac{\beta}{\eta}\right)\left(\frac{(x-\gamma)}{\eta}\right)^{\beta-1}$$

here, β, η, and γ are shape, scale, and location parameters, respectively, as in the three-parameter Weibull distribution.

Adding the two hazard functions gives the five-parameter bi-Weibull distribution, for which the hazard and reliability equations are:

Hazard function

$$h(x) = \begin{cases} \lambda\theta(\lambda x)^{\theta-1} & 0 < x < \gamma \\ \lambda\theta(\lambda x)^{\theta-1} + \left(\frac{\beta}{\eta}\right)\left(\frac{(x-\gamma)}{\eta}\right)^{\beta-1} & x \geq \gamma \end{cases}$$

Survival function

$$S(x) = \begin{cases} e^{-(\lambda x)^{\theta}} & 0 < x < \gamma \\ \exp(-\{(\lambda x)^{\theta} + [(x-\gamma)/\eta]^{\beta}\}) & x \geq \gamma \end{cases}$$

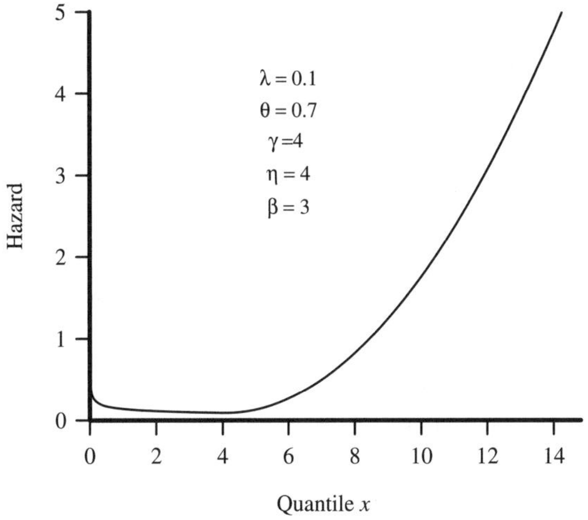

Figure 46.7. Bi-Weibull hazard function $\boldsymbol{W}$: $\lambda, \theta, \gamma, \eta, \beta$.

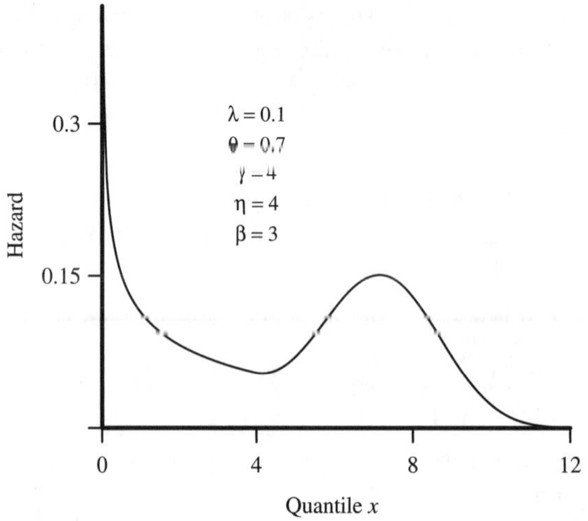

Figure 46.8. Bi-Weibull probability density function $\boldsymbol{W}$: $\lambda, \theta, \gamma, \eta, \ \beta$.

Bi-Weibull Random Number Generation

Use the equations given for the bi-Weibull survival function $S(x)$ to calculate values of $S(x)$, for values of x from 0 to $\gamma + 2\eta$, and keep the results in a table. Generate a uniform random number $\boldsymbol{R}$ and look up the value of x corresponding to $S(x) = \boldsymbol{R}$ (approximately) in the table.

Bi-Weibull Graphs

Figures 46.7, 46.8, 46.9 show, respectively, examples of the five-parameter bi-Weibull hazard function, probability density function, and reliability function. Note that the

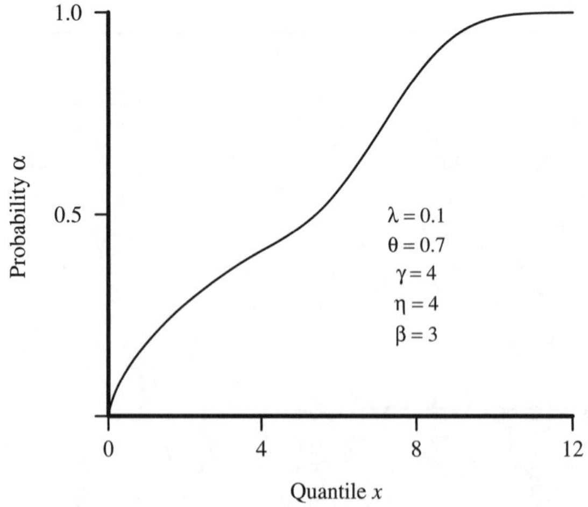

Figure 46.9. Bi-Weibull distribution function $\boldsymbol{W}$: $\lambda, \theta, \gamma, \eta, \ \beta$.

hazard function has a "bathtub" shape, which in a reliability application corresponds to a combination of burn-in and wearout failures. The range of shapes that can be taken by the bi-Weibull distribution is large. Any combination of two Weibull failure rate patterns can be accommodated, for example, burn-in plus wearout, random plus wearout, burn-in plus random, random plus another random starting later. β is not required to be greater than 1, nor θ less than 1. In practice, the use of the five-parameter bi-Weibull distribution to detect the onset of wearout is one of its main advantages.

46.9 WEIBULL FAMILY

The negative exponential, two-parameter Weibull, three-parameter Weibull, and bi-Weibull distributions form a family of distributions of gradually increasing complexity.

The negative exponential is the simplest and has a constant hazard function. The two-parameter Weibull extends the range of models to include (one of) decreasing, constant, or increasing hazard function. The three-parameter Weibull model adds a location parameter to the two-parameter model. The bi-Weibull distribution allows (two of) decreasing, constant, and increasing hazard function.

Chapter 47

Wishart (Central) Distribution

Matrix variate: $\boldsymbol{WC}$: $k, n, \boldsymbol{\Sigma}$.

Formed from n-independent multinormal variates $\boldsymbol{MN}$: $\boldsymbol{\mu}, \boldsymbol{\Sigma}$ by

$$\boldsymbol{WC}: k, n, \boldsymbol{\Sigma} \sim \sum_{i=1}^{n} (\boldsymbol{MN}: \boldsymbol{\mu}_i, \boldsymbol{\Sigma} - \boldsymbol{\mu}_i)(\boldsymbol{MN}: \boldsymbol{\mu}_i, \boldsymbol{\Sigma} - \boldsymbol{\mu}_i)'.$$

Matrix quantile $\boldsymbol{X}$ a $k \times k$ positive semidefinite matrix, with elements $\mathbf{X}_{ij}$.

Parameters $k, n \geq k, \boldsymbol{\Sigma}$, where k is the dimension of the n-associated multinormal multivariates; n is the degrees of freedom, $n \geq k$; and $\boldsymbol{\Sigma}$ is the $k \times k$ variance-covariance matrix of the associated multinormal multivariates, with elements $\Sigma_{ij} = \sigma_{ij}$.

Distribution function	$\dfrac{\Gamma_k[(k+1)/2]\lvert\boldsymbol{X}\rvert^{n/2}{}_1\boldsymbol{F}_1\left(n/2; (k+1)/2; -\frac{1}{2}\boldsymbol{\Sigma}^{-1}\boldsymbol{X}\right)}{\Gamma_k\left[\frac{1}{2}(n+k+1)\right]\lvert 2\boldsymbol{\Sigma}\rvert^{n/2}}$ where ${}_1F_1$ is a hypergeometric function of the matrix argument
Probability density function	$\exp\left(-\dfrac{1}{2}\mathrm{tr}\boldsymbol{\Sigma}^{-1}\boldsymbol{X}\right)\lvert\boldsymbol{X}\rvert^{(n-k-1)/2} \Big/ \left\{\Gamma_k\left(\dfrac{n}{2}\right)\lvert 2\boldsymbol{\Sigma}\rvert^{n/2}\right\}$
Characteristic function	$\lvert\boldsymbol{I}_k - 2i\boldsymbol{\Sigma}\boldsymbol{T}\rvert^{-n/2}$, where $\boldsymbol{T}$ is a symmetric $k \times k$ matrix such that $\boldsymbol{\Sigma}^{-1} - 2\boldsymbol{T}$ is positive define
Moment generating function	$\lvert\boldsymbol{I}_k - 2\boldsymbol{\Sigma}\boldsymbol{T}\rvert^{-n/2}$
rth Moment about origin	$\lvert 2\boldsymbol{\Sigma}\rvert^r\Gamma_k\left(\dfrac{1}{2}n + r\right) \Big/ \Gamma_k\left(\dfrac{1}{2}n\right)$
Mean	$n\boldsymbol{\Sigma}$
Individual elements	$E(\mathrm{X}_{ij}) = n\sigma_{ij}$ $\mathrm{cov}(\mathrm{X}_{ij}, \mathrm{X}_{rs}) = n(\sigma_{ir}\sigma_{js} + \sigma_{is}\sigma_{jr})$

Statistical Distributions, Fourth Edition, by Catherine Forbes, Merran Evans, Nicholas Hastings, and Brian Peacock

47.1 NOTE

The Wishart variate is a k-dimensional generalization of the chi-squared variate, which is the sum of squared normal variates. It performs a corresponding role for multivariate normal problems as the chi-squared does for the univariate normal.

47.2 VARIATE RELATIONSHIPS

1. The Wishart $k \times k$ matrix variate ***WC***: k, n, $\boldsymbol{\Sigma}$ is related to n-independent multinormal multivariates of dimension k, ***MN***: $\boldsymbol{\mu}, \boldsymbol{\Sigma}$, by

$$\boldsymbol{WC} : k, n, \boldsymbol{\Sigma} \sim \sum_{i=1}^{n} (\boldsymbol{MN} : \boldsymbol{\mu}_i, \boldsymbol{\Sigma} - \boldsymbol{\mu}_i)(\boldsymbol{MN} : \boldsymbol{\mu}_i, \boldsymbol{\Sigma} - \boldsymbol{\mu}_i)'.$$

2. The sum of mutually independent Wishart variates ***WC***: k, n_i, $\boldsymbol{\Sigma}$ is also a Wishart variate with parameters k, Σn_i, $\boldsymbol{\Sigma}$.

$$\sum (\boldsymbol{WC} : k, n_i, \boldsymbol{\Sigma}) \sim \boldsymbol{WC} : k, \sum n_i, \boldsymbol{\Sigma}.$$

Chapter 48

Statistical Tables

Statistical Distributions, Fourth Edition, by Catherine Forbes, Merran Evans, Nicholas Hastings, and Brian Peacock

Table 48.1: Normal Distribution Function – $F_N(x)$.

x	.00	.01	.02	.03	.04	.05	.06	.07	.08	.09
.0	0.5000	0.5040	0.5080	0.5120	0.5160	0.5199	0.5239	0.5279	0.5319	0.5359
.1	0.5398	0.5438	0.5478	0.5517	0.5557	0.5596	0.5636	0.5675	0.5714	0.5753
.2	0.5793	0.5832	0.5871	0.5910	0.5948	0.5987	0.6026	0.6064	0.6103	0.6141
.3	0.6179	0.6217	0.6255	0.6293	0.6331	0.6368	0.6406	0.6443	0.6480	0.6517
.4	0.6554	0.6591	0.6628	0.6664	0.6700	0.6736	0.6772	0.6808	0.6844	0.6879
.5	0.6915	0.6950	0.6985	0.7019	0.7054	0.7088	0.7123	0.7157	0.7190	0.7224
.6	0.7257	0.7291	0.7324	0.7357	0.7389	0.7422	0.7454	0.7486	0.7517	0.7549
.7	0.7580	0.7611	0.7642	0.7673	0.7704	0.7734	0.7764	0.7794	0.7823	0.7852
.8	0.7881	0.7910	0.7939	0.7967	0.7995	0.8023	0.8051	0.8078	0.8106	0.8133
.9	0.8159	0.8186	0.8212	0.8238	0.8264	0.8289	0.8315	0.8340	0.8365	0.8389
1	0.8413	0.8438	0.8461	0.8485	0.8508	0.8531	0.8554	0.8577	0.8599	0.8621
1.1	0.8643	0.8665	0.8686	0.8708	0.8729	0.8749	0.8770	0.8790	0.8810	0.8830
1.2	0.8849	0.8869	0.8888	0.8907	0.8925	0.8944	0.8962	0.8980	0.8997	0.9015
1.3	0.9032	0.9049	0.9066	0.9082	0.9099	0.9115	0.9131	0.9147	0.9162	0.9177
1.4	0.9192	0.9207	0.9222	0.9236	0.9251	0.9265	0.9279	0.9292	0.9306	0.9319
1.5	0.9332	0.9345	0.9357	0.9370	0.9382	0.9394	0.9406	0.9418	0.9429	0.9441
1.6	0.9452	0.9463	0.9474	0.9484	0.9495	0.9505	0.9515	0.9525	0.9535	0.9545
1.7	0.9554	0.9564	0.9573	0.9582	0.9591	0.9599	0.9608	0.9616	0.9625	0.9633
1.8	0.9641	0.9649	0.9656	0.9664	0.9671	0.9678	0.9686	0.9693	0.9699	0.9706
1.9	0.9713	0.9719	0.9726	0.9732	0.9738	0.9744	0.9750	0.9756	0.9761	0.9767
2	0.9772	0.9778	0.9783	0.9788	0.9793	0.9798	0.9803	0.9808	0.9812	0.9817
2.1	0.9821	0.9826	0.9830	0.9834	0.9838	0.9842	0.9846	0.9850	0.9854	0.9857
2.2	0.9861	0.9864	0.9868	0.9871	0.9875	0.9878	0.9881	0.9884	0.9887	0.9890
2.3	0.9893	0.9896	0.9898	0.9901	0.9904	0.9906	0.9909	0.9911	0.9913	0.9916
2.4	0.9918	0.9920	09922	0.9925	0.9927	0.9929	0.9931	0.9932	0.9934	0.9936
2.5	0.9938	0.9940	0.9941	0.9943	0.9945	0.9946	0.9948	0.9949	0.9951	0.9952
2.6	0.9953	0.9955	0.9956	0.9957	0.9959	0.9960	0.9961	0.9962	0.9963	0.9964
2.7	0.9965	0.9966	0.9967	0.9968	0.9969	0.9970	0.9971	0.9972	0.9973	0.9974
2.8	0.9974	0.9975	0.9976	0.9977	0.9977	0.9978	0.9979	0.9979	0.9980	0.9981
2.9	0.9981	0.9982	0.9982	0.9983	0.9984	0.9984	0.9985	0.9985	0.9986	0.9986
3	0.9987	0.9987	0.9987	0.9988	0.9988	0.9989	0.9989	0.9989	0.9990	0.9990
3.1	0.9990	0.9991	0.9991	0.9991	0.9992	0.9992	0.9992	0.9992	0.9993	0.9993
3.2	0.9993	0.9993	0.9994	0.9994	0.9994	0.9994	0.9994	0.9995	0.9995	0.9995
3.3	0.9995	0.9995	0.9995	0.9996	0.9996	0.9996	0.9996	0.9996	0.9996	0.9997
3.4	0.9997	0.9997	0.9997	0.9997	0.9997	0.9997	0.9997	0.9997	0.9997	0.9998

Note: $F_N(x) = \int_{-\infty}^{x} (2\pi)^{\frac{1}{2}} \exp\left\{-\frac{u^2}{2}\right\} du$. Reproduced with permission of the Biometrika Trustees from Pearson and Hartley (1966).

Table 48.2: Percentiles of the Chi-Squared $\chi^2 : \nu$ Distribution, $G(l - \alpha)$.

$\nu \backslash 1-\alpha$	0.99	0.975	0.95	0.9	0.1	0.05	0.025	0.01
1	1.57088E-04	9.82069E-04	3.93214E-03	0.015791	2.70554	3.84146	5.02389	6.63490
2	0.020101	0.0506356	0.102587	0.210721	4.60517	5.99146	7.37776	9.21034
3	0.114832	0.215795	0.351846	0.584374	6.25139	7.81473	9.34840	11.3449
4	0.297109	0.484419	0.710723	1.063623	7.77944	9.48773	11.1433	13.2767
5	0.554298	0.831212	1.145476	1.61031	9.23636	11.0705	12.8325	15.0863
6	0.872090	1.23734	1.63538	2.20413	10.6446	12.5916	14.4494	16.8119
7	1.239042	1.68987	2.16735	2.83311	12.0170	14.0671	16.0128	18.4753
8	1.64650	2.17973	2.73264	3.48954	13.3616	15.5073	17.5345	20.0902
9	2.08790	2.70039	3.32511	4.16816	14.6837	16.9190	19.0228	21.6660
10	2.55821	3.24697	3.94030	4.86518	15.9872	18.3070	20.4832	23.2093
11	3.05348	3.81575	4.57481	5.57778	17.2750	19.6751	21.9200	24.7250
12	3.57057	4.40379	5.22603	6.30380	18.5493	21.0261	23.3367	26.2170
13	4.10692	5.00875	5.89186	7.04150	19.8119	22.3620	24.7356	27.6882
14	4.66043	5.62873	6.57063	7.78953	21.0641	23.6848	26.1189	29.1412
15	5.22935	6.26214	7.26094	8.54676	22.3071	24.9958	27.4884	30.5779
16	5.81221	6.90766	7.96165	9.31224	23.5418	26.2962	28.8454	31.9999
17	6.40776	7.56419	8.67176	10.0852	24.7690	27.5871	30.1910	33.4087
18	7.01491	8.23075	9.39046	10.8649	25.9894	28.8693	31.5264	34.8053
19	7.63273	8.90652	10.1170	11.6509	27.2036	30.1435	32.8523	36.1909
20	8.26040	9.59078	10.8508	12.4426	28.4120	31.4104	34.1696	37.5662
21	8.89720	10.2829	11.5913	13.2396	29.6151	32.6706	35.4789	38.9322
22	9.54249	10.9823	12.3380	14.0415	30.8133	33.9244	36.7807	40.2894
23	10.1957	11.6886	13.0905	14.8480	32.0069	35.1725	38.0756	41.6384
24	10.8564	12.4012	13.8484	15.6587	33.1962	36.4150	39.3641	42.9798
25	11.5240	13.1197	14.6114	16.4734	34.3816	37.6525	40.6465	44.3141
26	12.1981	13.8439	15.3792	17.2919	35.5632	38.8851	41.9232	45.6417
27	12.8785	14.5734	16.1514	18.1139	36.7412	40.1133	43.1945	46.9629
28	13.5647	15.3079	16.9279	18.9392	37.9159	41.3371	44.4608	48.2782
29	14.2565	16.0471	17.7084	19.7677	39.0875	42.5570	45.7223	49.5879
30	14.9535	16.7908	18.4927	20.5992	40.2560	43.7730	46.9792	50.8922
40	22.1643	24.4330	26.5093	29.0505	51.8051	55.7585	59.3417	63.6907
50	29.7067	32.3574	34.7643	37.6886	63.1671	67.5048	71.4202	76.1539
60	37.4849	40.4817	43.1880	46.4589	74.3970	79.0819	83.2977	88.3794
70	45.4417	48.7576	51.7393	55.3289	85.5270	90.5312	95.0232	100.425
80	53.5401	57.1532	60.3915	64.2778	96.5782	101.879	106.629	112.329
90	61.7541	65.6466	69.1260	73.2911	107.565	113.145	118.136	124.116
100	70.0649	74.2219	77.9295	82.3581	118.498	124.342	129.561	135.807

Note: $\Pr[\chi^2 : \nu > G(l - \alpha)] = \alpha$. Reproduced with permission of the Biometrika Trustees from Pearson and Hartley (1966).

Table 48.3: Percentiles of the $F : \nu, \omega$ Distribution

Upper 5% points = $G(0.95)$

ω \ ν	1	2	3	4	5	6	7	8	9	10	12	15	20	24	30	40	60	120	∞
1	161.4	199.5	215.7	224.6	230.2	234.0	236.8	238.9	240.5	241.9	243.9	245.9	248.0	249.1	250.1	251.1	252.2	253.3	254.3
2	18.51	19.00	19.16	19.25	19.30	19.33	19.35	19.37	19.38	19.40	19.41	19.43	19.45	19.45	19.46	19.47	19.48	19.49	19.50
3	10.13	9.55	9.28	9.12	9.01	8.94	8.89	8.85	8.81	8.79	8.74	8.70	8.66	8.64	8.62	8.59	8.57	8.55	8.53
4	7.71	6.94	6.59	6.39	6.26	6.16	6.09	6.04	6.00	5.96	5.91	5.86	5.80	5.77	5.75	5.72	5.69	5.66	5.63
5	6.61	5.79	5.41	5.19	5.05	4.95	4.88	4.82	4.77	4.74	4.68	4.62	4.56	4.53	4.50	4.46	4.43	4.40	4.36
6	5.99	5.14	4.76	4.53	4.39	4.28	4.21	4.15	4.10	4.06	4.00	3.94	3.87	3.84	3.81	3.77	3.74	3.70	3.67
7	5.59	4.74	4.35	4.12	3.97	3.87	3.79	3.73	3.68	3.64	3.57	3.51	3.44	3.41	3.38	3.34	3.30	3.27	3.23
8	5.32	4.46	4.07	3.84	3.69	3.58	3.50	3.44	3.39	3.35	3.28	3.22	3.15	3.12	3.08	3.04	3.01	2.97	2.93
9	5.12	4.26	3.86	3.63	3.48	3.37	3.29	3.23	3.18	3.14	3.07	3.01	2.94	2.90	2.86	2.83	2.79	2.75	2.71
10	4.96	4.10	3.71	3.48	3.33	3.22	3.14	3.07	3.02	2.98	2.91	2.85	2.77	2.74	2.70	2.66	2.62	2.58	2.54
11	4.84	3.98	3.59	3.36	3.20	3.09	3.01	2.95	2.90	2.85	2.79	2.72	2.65	2.61	2.57	2.53	2.49	2.45	2.40
12	4.75	3.89	3.49	3.26	3.11	3.00	2.91	2.85	2.80	2.75	2.69	2.62	2.54	2.51	2.47	2.43	2.38	2.34	2.30
13	4.67	3.81	3.41	3.18	3.03	2.92	2.83	2.77	2.71	2.67	2.60	2.53	2.46	2.42	2.38	2.34	2.30	2.25	2.21
14	4.60	3.74	3.34	3.11	2.96	2.85	2.76	2.70	2.65	2.60	2.53	2.46	2.39	2.35	2.31	2.27	2.22	2.18	2.13
15	4.54	3.68	3.29	3.06	2.90	2.79	2.71	2.64	2.59	2.54	2.48	2.40	2.33	2.29	2.25	2.20	2.16	2.11	2.07
16	4.49	3.63	3.24	3.01	2.85	2.74	2.66	2.59	2.54	2.49	2.42	2.35	2.28	2.24	2.19	2.15	2.11	2.06	2.01
17	4.45	3.59	3.20	2.96	2.81	2.70	2.61	2.55	2.49	2.45	2.38	2.31	2.23	2.19	2.15	2.10	2.06	2.01	1.96
18	4.41	3.55	3.16	2.93	2.77	2.66	2.58	2.51	2.46	2.41	2.34	2.27	2.19	2.15	2.11	2.06	2.02	1.97	1.92
19	4.38	3.52	3.13	2.90	2.74	2.63	2.54	2.48	2.42	2.38	2.31	2.23	2.16	2.11	2.07	2.03	1.98	1.93	1.88
20	4.35	3.49	3.10	2.87	2.71	2.60	2.51	2.45	2.39	2.35	2.28	2.20	2.12	2.08	2.04	1.99	1.95	1.90	1.84
21	4.32	3.47	3.07	2.84	2.68	2.57	2.49	2.42	2.37	2.32	2.25	2.18	2.10	2.05	2.01	1.96	1.92	1.87	1.81
22	4.30	3.44	3.05	2.82	2.66	2.55	2.46	2.40	2.34	2.30	2.23	2.15	2.07	2.03	1.98	1.94	1.89	1.84	1.78
23	4.28	3.42	3.03	2.80	2.64	2.53	2.44	2.37	2.32	2.27	2.20	2.13	2.05	2.01	1.96	1.91	1.86	1.81	1.76
24	4.26	3.40	3.01	2.78	2.62	2.51	2.42	2.36	2.30	2.25	2.18	2.11	2.03	1.98	1.94	1.89	1.84	1.79	1.73
25	4.24	3.39	2.99	2.76	2.60	2.49	2.40	2.34	2.28	2.24	2.16	2.09	2.01	1.96	1.92	1.87	1.82	1.77	1.71
26	4.23	3.37	2.98	2.74	2.59	2.47	2.39	2.32	2.27	2.22	2.15	2.07	1.99	1.95	1.90	1.85	1.80	1.75	1.69
27	4.21	3.35	2.96	2.73	2.57	2.46	2.37	2.31	2.25	2.20	2.13	2.06	1.97	1.93	1.88	1.84	1.79	1.73	1.67
28	4.20	3.34	2.95	2.71	2.56	2.45	2.36	2.29	2.24	2.19	2.12	2.04	1.96	1.91	1.87	1.82	1.77	1.71	1.65
29	4.18	3.33	2.93	2.70	2.55	2.43	2.35	2.28	2.22	2.18	2.10	2.03	1.94	1.90	1.85	1.81	1.75	1.70	1.64
30	4.17	3.32	2.92	2.69	2.53	2.42	2.33	2.27	2.21	2.16	2.09	2.01	1.93	1.89	1.84	1.79	1.74	1.68	1.62
40	408	3.23	2.84	2.61	2.45	234	2.25	2.18	2.12	2.08	2.00	1.92	1.84	1.79	1.74	1.69	1.64	1.58	1.51
60	4.00	3.15	2.76	2.53	2.37	2.25	2.17	2.10	2.04	1.99	1.92	1.84	1.75	1.70	1.65	1.59	1.53	1.47	1.39
120	3.92	3.07	2.68	2.45	2.29	2.17	2.09	2.02	1.96	1.91	1.83	1.75	1.66	1.61	1.55	1.50	1.43	1.35	1.25
∞	3.84	3.00	2.60	2.37	2.21	2.10	2.01	1.94	1.88	1.83	1.75	1.67	1.57	1.52	1.46	1.39	1.32	1.22	1.01

Table 48.3: (*Continued*)

Upper 1% points = $G(0.99)$

$\omega \backslash \nu$	1	2	3	4	5	6	7	8	9	10	12	15	20	24	30	40	60	120	∞
1	4052	4999.5	5403	5625	5764	5859	5928	5981	6022	6056	6106	6157	6209	6235	6261	6287	6313	6339	6366
2	98.50	99.00	99.17	99.25	99.30	99.33	99.36	99.37	99.39	99.40	99.42	99.43	99.45	99.46	99.47	99.47	99.48	99.49	99.50
3	34.12	30.82	29.46	28.71	28.24	27.91	27.67	27.49	27.35	27.23	27.05	26.87	26.69	26.60	26.50	26.41	26.32	26.22	26.13
4	21.20	18.00	16.69	15.98	15.52	15.21	14.98	14.80	14.66	14.55	14.37	14.20	14.02	13.93	13.84	13.75	13.65	13.56	13.46
5	16.26	13.27	12.06	11.39	10.97	10.67	10.46	10.29	10.16	10.05	9.89	9.72	9.55	9.47	9.38	9.29	9.20	9.11	9.02
6	13.75	10.92	9.78	9.15	8.75	8.47	8.26	8.10	7.98	7.87	7.72	7.56	7.40	7.31	7.23	7.14	7.06	6.97	6.88
7	12.25	9.55	8.45	7.85	7.46	7.19	6.99	6.84	6.72	6.62	6.47	6.31	6.16	6.07	5.99	5.91	5.82	5.74	5.65
8	11.26	8.65	7.59	7.01	6.63	6.37	6.18	6.03	5.91	5.81	5.67	5.52	5.36	5.28	5.20	5.12	5.03	4.95	4.86
9	10.56	8.02	6.99	6.42	6.06	5.80	5.61	5.47	5.35	5.26	5.11	4.96	4.81	4.73	4.65	4.57	4.48	4.40	4.31
10	10.04	7.56	6.55	5.99	5.64	5.39	5.20	5.06	4.94	4.85	4.71	4.56	4.41	4.33	4.25	4.17	4.08	4.00	3.91
11	9.65	7.21	6.22	5.67	5.32	5.07	4.89	4.74	4.63	4.54	4.40	4.25	4.10	4.02	3.94	3.86	3.78	3.69	3.60
12	9.33	6.93	5.95	5.41	5.06	4.82	4.64	4.50	4.39	4.30	4.16	4.01	3.86	3.78	3.70	3.62	3.54	3.45	3.36
13	9.07	6.70	5.74	5.21	4.86	4.62	4.44	4.30	4.19	4.10	3.96	3.82	3.66	3.59	3.51	3.43	3.34	3.25	3.17
14	8.86	6.51	5.56	5.04	4.69	4.46	4.28	4.14	4.03	3.94	3.80	3.66	3.51	3.43	3.35	3.27	3.18	3.09	3.00
15	8.68	6.36	5.42	4.89	4.56	4.32	4.14	4.00	3.89	3.80	3.67	3.52	3.37	3.29	3.21	3.13	3.05	2.96	2.87
16	8.53	6.23	5.29	4.77	4.44	4.20	4.03	3.89	3.78	3.69	3.55	3.41	3.26	3.18	3.10	3.02	2.93	2.84	2.75
17	8.40	6.11	5.18	4.67	4.34	4.10	3.93	3.79	3.68	3.59	3.46	3.31	3.16	3.08	3.00	2.92	2.83	2.75	2.65
18	8.29	6.01	5.09	4.58	4.25	4.01	3.84	3.71	3.60	3.51	3.37	3.23	3.08	3.00	2.92	2.84	2.75	2.66	2.57
19	8.18	5.93	5.01	4.50	4.17	3.94	3.77	3.63	3.52	3.43	3.30	3.15	3.00	2.92	2.84	2.76	2.67	2.58	2.49
20	8.10	5.85	4.94	4.43	4.10	3.87	3.70	3.56	3.46	3.37	3.23	3.09	2.94	2.86	2.78	2.69	2.61	2.52	2.42
21	8.02	5.78	4.87	4.37	4.04	3.81	3.64	3.51	3.40	3.31	3.17	3.03	2.88	2.80	2.72	2.64	2.55	2.46	2.36
22	7.95	5.72	4.82	4.31	3.99	3.76	3.59	3.45	3.35	3.26	3.12	2.98	2.83	2.75	2.67	2.58	2.50	2.40	2.31
23	7.88	5.66	4.76	4.26	3.94	3.71	3.54	3.41	3.30	3.21	3.07	2.93	2.78	2.70	2.62	2.54	2.45	2.35	2.26
24	7.82	5.61	4.72	4.22	3.90	3.67	3.50	3.36	3.26	3.17	3.03	2.89	2.74	2.66	2.58	2.49	2.40	2.31	2.21
25	7.77	5.57	4.68	4.18	3.85	3.63	3.46	3.32	3.22	3.13	2.99	2.85	2.70	2.62	2.54	2.45	2.36	2.27	2.17
26	7.72	5.53	4.64	4.14	3.82	3.59	3.42	3.29	3.18	3.09	2.96	2.81	2.66	2.58	2.50	2.42	2.33	2.23	2.13
27	7.68	5.49	4.60	4.11	3.78	3.56	3.39	3.26	3.15	3.06	2.93	2.78	2.63	2.55	2.47	2.38	2.29	2.20	2.10
28	7.64	5.45	4.57	4.07	3.75	3.53	3.36	3.23	3.12	3.03	2.90	2.75	2.60	2.52	2.44	2.35	2.26	2.17	2.06
29	7.60	5.42	4.54	4.04	3.73	3.50	3.33	3.20	3.09	3.00	287	2.73	2.57	2.49	2.41	2.33	2.23	2.14	2.03
30	7.56	5.39	4.51	4.02	3.70	3.47	3.30	3.17	3.07	2.98	2.84	2.70	2.55	2.47	2.39	2.30	2.21	2.11	2.01
40	7.31	5.18	4.31	3.83	3.51	3.29	3.12	2.99	2.89	2.80	2.66	2.52	2.37	2.29	2.20	2.11	2.02	1.92	1.80
60	7.08	4.98	4.13	3.65	3.34	3.12	2.95	2.82	2.72	2.63	2.50	2.35	2.20	2.12	2.03	1.94	1.84	1.73	1.60
120	6.85	4.79	3.95	3.48	3.17	2.96	2.79	2.66	2.56	2.47	2.34	2.19	2.03	1.95	1.86	1.76	1.66	1.53	1.38
∞	6.64	4.61	3.78	3.32	3.02	2.80	2.64	2.51	2.41	2.32	2.18	2.04	1.88	1.79	1.70	1.59	1.47	1.32	1.00

Note: $\Pr[F : \nu, \omega > G(1-\alpha)] = \alpha$. Reproduced with permission of the Biometrika Trustees from Pearson and Hartley (1966).

Table 48.4: Percentiles of the Student's *t* Distribution

ν	$(1-\alpha)=0.4$	0.25	0.1	0.05	0.025	0.01	0.005	0.0025	0.001	0.0005
	$(1-2\alpha)=0.8$	0.5	0.2	0.1	0.05	0.02	0.01	0.005	0.002	0.001
1	0.325	1.000	3.078	6.314	12.706	31.821	63.657	127.32	318.31	636.619
2	0.289	0.816	1.886	2.920	4.303	6.965	9.925	14.089	22.327	31.599
3	0.277	0.765	1.638	2.353	3.182	4.541	5.841	7.453	10.214	12.924
4	0.271	0.741	1.533	2.132	2.776	3.747	4.604	5.598	7.173	8.610
5	0.267	0.727	1.476	2.015	2.571	3.365	4.032	4.773	5.893	6.869
6	0.265	0.718	1.440	1.943	2.447	3.143	3.707	4.317	5.208	5.959
7	0.263	0.711	1.415	1.895	2.365	2.998	3.499	4.029	4.785	5.408
8	0.262	0.706	1.397	1.860	2.306	2.896	3.355	3.833	4.501	5.041
9	0.261	0.703	1.383	1.833	2.262	2.821	3.250	3.690	4.297	4.781
10	0.260	0.700	1.372	1.812	2.228	2.764	3.169	3.581	4.144	4.587
11	0.260	0.697	1.363	1.796	2.201	2.718	3.106	3.497	4.025	4.437
12	0.259	0.695	1.356	1.782	2.179	2.681	3.055	3.428	3.930	4.318
13	0.259	0.694	1.350	1.771	2.160	2.650	3.012	3.372	3.852	4.221
14	0.258	0.692	1.345	1.761	2.145	2.624	2.977	3.326	3.787	4.140
15	0.258	0.691	1.341	1.753	2.131	2.602	2.947	3.286	3.733	4.073
16	0.258	0.690	1.337	1.746	2.120	2.583	2.921	3.252	3.686	4.015
17	0.257	0.689	1.333	1.740	2.110	2.567	2.898	3.222	3.646	3.965
18	0.257	0.688	1.330	1.734	2.101	2.552	2.878	3.197	3.610	3.922
19	0.257	0.688	1.328	1.729	2.093	2.539	2.861	3.174	3.579	3.883
20	0.257	0.687	1.325	1.725	2.086	2.528	2.845	3.153	3.552	3.850
21	0.257	0.686	1.323	1.721	2.080	2.518	2.831	3.135	3.527	3.819
22	0.256	0.686	1.321	1.717	2.074	2.508	2.819	3.119	3.505	3.792
23	0.256	0.685	1.319	1.714	2.069	2.500	2.807	3.104	3.485	3.767
24	0.256	0.685	1.318	1.711	2.064	2.492	2.797	3.091	3.467	3.745
25	0.256	0.684	1.316	1.708	2.060	2.485	2.787	3.078	3.450	3.725
26	0.256	0.684	1.315	1.706	2.056	2.479	2.779	3.067	3.435	3.707
27	0.256	0.684	1.314	1.703	2.052	2.473	2.771	3.057	3.421	3.690
28	0.256	0.683	1.313	1.701	2.048	2.467	2.763	3.047	3.408	3.674
29	0.256	0.683	1.311	1.699	2.045	2.462	2.756	3.038	3.396	3.659
30	0.256	0.683	1.310	1.697	2.042	2.457	2.750	3.030	3.385	3.646
40	0.255	0.681	1.303	1.684	2.021	2.423	2.704	2.971	3.307	3.551
60	0.254	0.679	1.296	1.671	2.000	2.390	2.660	2.915	3.232	3.460
120	0.254	0.677	1.289	1.658	1.980	2.358	2.617	2.860	3.160	3.373
∞	0.253	0.674	1.282	1.645	1.960	2.326	2.576	2.807	3.090	3.291

Note: $1-\alpha$ is the upper-tail area of the distribution for v degrees of freedom, appropriate for use in a single-tail test. For a two-tail test, $1-2\alpha$ must be used. Reproduced with permission of the Biometrika Trustees from Pearson and Hartley (1966).

Table 48.5: Partial Expectations for the Standard Normal Distribution

Partial expectation $P(x) = \int_x^{\infty} (u - x) f(u)\, du$
When $x < 3.0$, use $-x$ as an approximation for the partial expectation

x	$P(x)$	x	$P(x)$	x	$P(x)$
−2.9	2.9005	−0.9	1.0004	1.1	0.0686
−2.8	2.8008	−0.8	0.9202	1.2	0.0561
−2.7	2.7011	−0.7	0.8429	1.3	0.0455
−2.6	2.6015	−0.6	0.7687	1.4	0.0367
2.5	2.5020	0.5	0.6987	1.5	0.0293
−2.4	2.4027	−0.4	0.6304	1.6	0.0232
−2.3	2.3037	−0.3	0.5668	1.7	0.0183
−2.2	2.2049	−0.2	0.5069	1.8	0.0143
−2.1	2.1065	−0.1	0.4509	1.9	0.0111
−2.0	2.0085	0.0	0.3989	2.0	0.0085
−1.9	1.9111	0.1	0.3509	2.1	0.0065
−1.8	1.8143	0.2	0.3069	2.2	0.0049
−1.7	1.7183	0.3	0.2668	2.3	0.0037
−1.6	1.6232	0.4	0.2304	2.4	0.0027
−1.5	1.5293	0.5	0.1978	2.5	0.0020
−1.4	1.4367	0.6	0.1687	2.6	0.0015
−1.3	1.3455	0.7	0.1429	2.7	0.0011
−1.2	1.2561	0.8	0.1202	2.8	0.0008
−1.1	1.1686	0.9	0.1004	2.9	0.0005
−1.0	1.0833	1.0	0.0833	3.0	0.0004

Bibliography

AZZALINI, A., A Class of Distributions Which Includes the Normal Ones, *Scandinavian Journal of Statistics*, **12**, 2, 171–178 (1985).

BARTLETT, H. J. G., and HASTINGS, N. A. J., Improved Reliability Estimates for Samples Containing Suspensions, *Journal of Quality in Maintenance Engineering*, **4**, 107–114 (1998).

BOX, G. E. P., JENKINS, G. M., and REINSEL, G. C., *Time Series Analysis: Forecasting and Control*, 4th ed., Wiley (2008).

CASELLA, G., and BERGER, R. L., *Statistical Inference*, 2nd ed., Thomson, (2002).

COOPER, R., *Introduction to Queuing Theory*, 2nd ed., North-Holland, (1981).

COX, D. R., *Principles of Statistical Inference*, Cambridge, (2006).

D'AGOSTINO, R. B., and STEPHENS, M. A., *Goodness of Fit Techniques*, Marcel Dekker (1986).

FERNÁNDEZ, C., and STEEL, M. F. J., On Bayesian Modeling of Fat Tails and Skewness, *Journal of the American Statistical Association*, **93**, 359–371 (1998).

FRÜHWIRTH-SCHNATTER, S., *Finite Mixture and Markov Switching Models*, Springer (2006).

HALL, A. R., *Generalized Method of Moments*, Oxford University Press (2005).

HAMILTON, J. D., *Time Series Analysis*, Princeton University Press (1994).

JOHNSON, N. L., and KOTZ, S. *Distributions in Statistics: Continuous Multivariate Distributions*, Wiley (1972).

JOHNSON, N. L., and KOTZ, S., *Urn Models and Their Application, an Approach to Modern Discrete Probability Theory*, Wiley (1977).

JOHNSON, N. L., and KOTZ, S., Developments in Discrete Distributions, 1969–1980. *International Statistical Review*, **50**, 70–101 (1982).

JOHNSON, N. L., KOTZ, S., and BALAKRISHNAN, N., *Continuous Univariate Distributions*, 2nd ed. (2 vols.), Wiley (1994).

JOHNSON, N. L., KOTZ, S., and KEMP, A. W., *Univariate Discrete Distributions*, 3rd ed., Wiley (2005).

JOHNSTON, J., DINARDO, J., *Econometric Methods*, 4th ed., McGraw-Hill (1997).

KENDALL, D., Some Problems in the Theory of Queues, *Journal of the Royal Statistical Society Series B*, **13**, 151–185, (1951).

KOTZ, S., and JOHNSON, N. L., *Encyclopedia of Statistical Sciences*, Volumes 1–9 and supplement, Wiley (1982–1989).

LEE, A., *Applied Queuing Theory*, St Martins Press, (1966).

LYE, J., and MARTIN, V. L., Robust Estimation, Nonnormalities, and Generalized Exponential Distributions, *Journal of the American Statistical Association*, **88**, No. 421, 261–267 (1993).

MARDIA, K. V., *Statistics of Directional Data*, Academic Press (1972).

MARRIOT, F. H. C., *A Dictionary of Statistical Terms*, 5th ed., Wiley, (1990).

Statistical Distributions, Fourth Edition, by Catherine Forbes, Merran Evans, Nicholas Hastings, and Brian Peacock

McCullagh, P., and Nelder, J. A., *Generalised Linear Models*, Chapman and Hall (1989).

Nadarajah, A. and Kotz, S., Skewed Distributions Generated by the Normal Kernal, *Statistics and Probability Letters*, **65**, 269–277 (2003).

O'Hagan, A. and Leonard, T., Bayes Estimation Subject to Uncertainty About Parameter Constraints, *Biometrika*, **63**, 2, 201–203 (1976).

Patel, J. K., Kapadia, C. H., and Owen, D. B., *Handbook of Statistical Distributions*, Marcel Dekker (1976).

Patil, G. P., Boswell, M. T., and Ratnaparkhi, M. V., *Dictionary and Classified Bibliography of Statistical Distributions in Scientific Work*: Vol. 1. *Discrete Models* (with S. W. Joshi); Vol. 2. *Continuous Univariate Models*; Vol. 3. *Multivariate Models* (with J. J. J. Roux), International Cooperative Publishing House (1984).

Pearson, E. S., and Hartley, H. O., eds., *Biometrika Tables for Statisticians*, Vol. 1, 3rd ed., Cambridge University Press (1966).

Pearson, K. ed., *Tables of Incomplete Beta Functions*, 2nd ed., Cambridge University Press (1968).

Robert, C. P., *The Bayesian Choice: From Decision-Theoretic Foundations to Computational Implementation*, 2nd ed., Springer (2007).

Stuart, A., and Ord, J. K., *Kendall's Advanced Theory of Statistics*, Vol. I, *Distribution Theory*, 6th ed., Hodder Arnold (1994).

Winston, W.L., *Operations Research*, Brooks/Cole-Thompson Learning, (2004).